DIGITAL SPECTRAL ANALYSIS

MATLAB® Software User Guide

S. LAWRENCE MARPLE JR.

Dover Publications, Inc.
Mineola, New York

Copyright

Bibliographical Note

Digital Spectral Analysis MATLAB Software User Guide is a new work, first published by Dover Publications, Inc. in 2019, and is the companion book to *Digital Spectral Analysis, Second Edition* (ISBN-13: 978-0-486-78052-8).

Library of Congress Cataloging-in-Publication Data

Names: Marple, S. Lawrence, Jr., author.
Title: Digital spectral analysis : MATLAB software user guide / S. Lawrence Marple, Jr.
Description: Second edition. | Mineola, New York : Dover Publications, Inc., 2019.
Identifiers: LCCN 2019002265| ISBN 9780486837383 | ISBN 0486837386
Subjects: LCSH: Spectral theory (Mathematics)—Handbooks, manuals, etc. | Signal processing—Digital techniques—Handbooks, manuals, etc. | MATLAB—Handbooks, manuals, etc.
Classification: LCC QA280 .M38 2019 Suppl. | DDC 519.5/502855133—dc23
LC record available at https://lccn.loc.gov/2019002265

Manufactured in the United States by LSC Communications
83738601 2019
www.doverpublications.com

Contents

Chapter 1

INTRODUCTION

This user guide is intended to be a companion to the textbook *Digital Spectral Analysis, Second Edition* (Dover Publications, 2019), illustrating all the techniques and algorithms described in the textbook. All references to the textbook throughout this user guide will simply use *Digital Spectral Analysis* for brevity. The spectral demonstrations use MATLAB software that encompass the full experience from inputting signal sources, interactively setting technique parameters and processing with those parameters, and choosing from a variety of plotting techniques to display the results. The processing functions and scripts have been coded to automatically handle sample data that is either real-valued or complex-valued. There are four software categories that support the demonstrations. These are the main MATLAB spectral demonstration scripts, supporting MATLAB plotting scripts, MATLAB technique processing functions, and signal sample data sources. The eleven MATLAB spectral demonstration scripts, each with the root name `spectrum_demo_xxx.m`, are provided in a zipped file on the Dover Publications website for *Digital Spectral Analysis* and this user guide. The seven plotting scripts, each with the root name `plot_xxx.m`, and the nine signal data sources are furnished in zipped files on the same website. The forty-four MATLAB functions are listed alphabetically in Chapter 6 of this user guide, together with input/output parameter characterizations and a brief processing description of each function.

Each demonstration script follows the same five-step organizational approach for consistency:

- SELECT AND IMPORT SIGNAL SOURCE. Scripts `spectrum_demo_xxx` have comment lines that show where a user may also import their own data.

- SELECT SHORT SIGNAL OR LONG SIGNAL ANALYSIS. Short signal analysis provides a one-dimensional frequency-only spectral plot result. Long signal analysis, with parameter choices, provides a two-dimensional time versus frequency spectral image result. This step is omitted for multi-channel and 2-D signal choices.

- SPECTRAL TECHNIQUE ALGORITHM AND PARAMETER SELECTIONS. Some spectral estimation techniques have more than one estimation algorithm from which to select.
- PROCESS SOURCE ACCORDING TO SPECTRAL TECHNIQUE SELECTION
- PLOT THE RESULTS. A variety of plotting choices is provided.

Please note that if you modify a demonstration script to also import your own data as a source, the input data must be in a MATLAB column vector of N samples if a single data source, or as an $N \times M$ MATLAB array for multichannel analysis, in which N samples are down rows and the M signal channels are across columns.

Demo Data Sources: Computer Generated

The nine data sources contain the four sources cited in *Digital Spectral Analysis* plus five additional sources for analysis demonstration in this user guide. All four *Digital Spectral Analysis*-cited sources are text-readable files with `*.dat` extensions. The computer-created data sources `test1987` and `doppler_radar` are computer-generated *short duration* data records with sixty-four complex-valued samples each at sampling rates of 1 and 2,500 samples/sec, respectively. MATLAB script `plot_text_data` may be used to plot these signals, as shown on right side of Figure 1.1. The short duration simulated doppler radar signal consists of complex sinusoidals (frequencies −750 Hz, −250 Hz, 500 Hz, 525 Hz, and 1,000 Hz at varying power levels) in additive colored noise that emulates

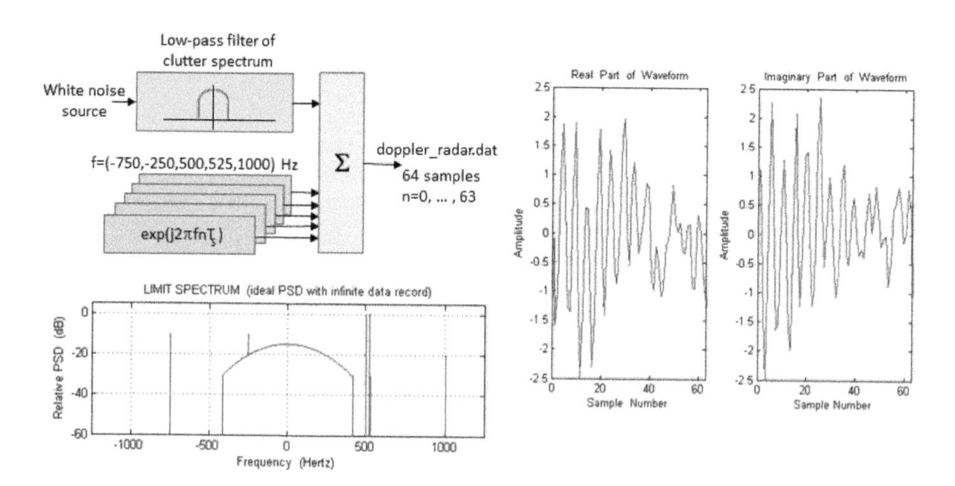

Figure 1.1: Plot of real and imaginary parts of `doppler_radar.dat` (right), signal generation scheme (upper left), and ideal spectrum (lower left).

clutter noise as indicated top left in Fig. 1.1. The ideal spectrum is analytically computed and plotted at lower left Fig. 1.1.

Demo Data Sources: Actual

The St. Louis temperature data and sunspot numbers (files `temperature` and `sunspot_numbers` cited in *Digital Spectral Analysis*) contain over one hundred overlapping years of once a month samples of monthly average temperatures and monthly average sunspot numbers. This signal pair will be used to illustrate two-channel spectral analysis (see Chapter 4 of this user guide).

Three *long duration* signal records have been added for this user guide to illustrate nonstationary signal application of techniques developed in *Digital Spectral Analysis* but not discussed in the text. These are depicted in Fig. 1.2. Actual signal in `bat_ultrasonic_pulse.mat` are samples of a single ultrasonic pulse [Fig. 1.2(a)] from a brown bat, which may be obtained from the website `www.ece.rice.edu/dsp/software`. Each pulse is used by the bat to echo locate surroundings and insects in flight. The pulse signal is a real-valued data record of four hundred samples at a sampling interval of seven microseconds (143 Ksps). Next, a 16,384 sample data record of ultrasonic cardiac features

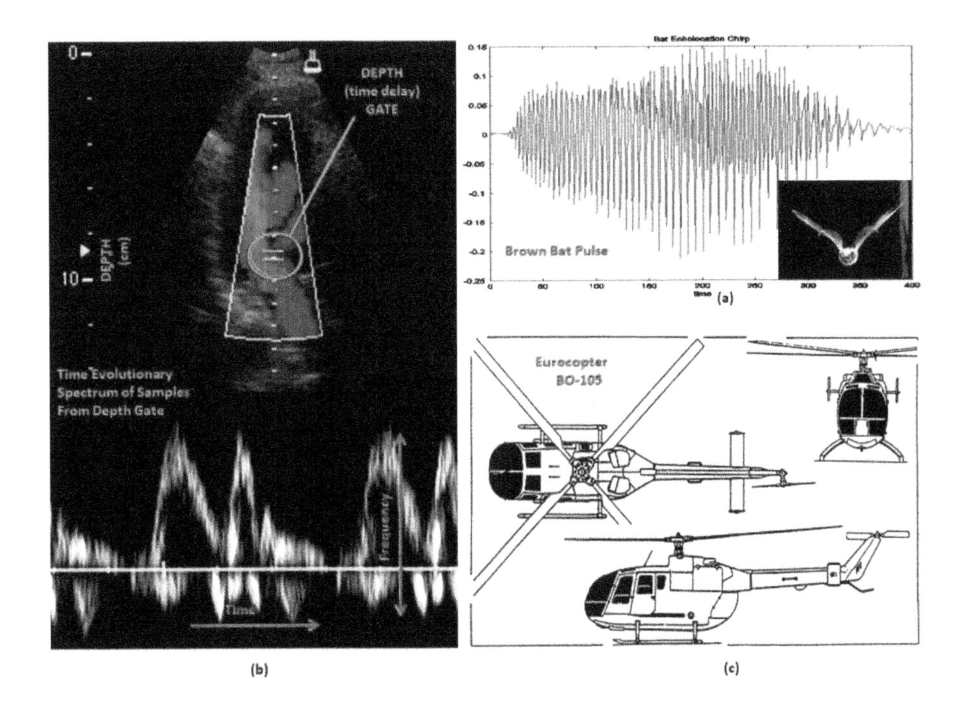

Figure 1.2: Signal sources: (a) Brown bat ultrasonic pulse. (b) Doppler ultrasound signature of blood flow in heart. (c) Helicopter in flight radar target.

(blood flow velocity) is provided in `doppler_heart.mat`. These features are used in a clinical diagnosis of heart health by examining how the blood flow rate changes during a beat cycle. The cardiac performance is made by visual evaluation of intracardiac velocity flow patterns. A physician seeks to detect flow disturbance (jet patterns, holes for fluid flow where it should not be, regurgitation due to blockage) due to defects using the time vs frequency (velocity) pattern. A medical ultrasound instrument launches pulses from a 5 MHz probe directed to a point in the heart (such as a valve), in this case at a pulse rate of 6,250 pulses per second. Based on the return time of the propagating ultrasound pulse relative to a specific depth, a time gate [shown set for a 9 cm depth in Fig. 1.2(b)] is used to capture a single complex-valued sample (due to complex demodulation from the probe frequency) with each pulse. An FFT of a moving analysis window on the collected data is then used to produce a sliding window spectral analysis of blood flow doppler [see bottom of Fig. 1.2(b)] to show change in blood flow velocity during a heart beat. The relationship between spectral frequency and blood flow rate in cm/sec is given by $v = \lambda f_{doppler}/2$ in which λ is the wavelength of the propagating ultrasound pulse [propagation rate varies with organ tissue encountered, but nominally is $c = 1540$ meters/sec].

Finally, a 4,096 sample data record of doppler radar features from an X-band radar illuminating a helicopter in flight is provided in `doppler_helicopter.mat`. The analysis to be illustrated in Chapter 3 will not only show the main radial velocity constant doppler component of the helicopter fuselage moving toward the radar, but also micro doppler variations of the other moving components on the helicopter (main rotor of four blades, tail rotor of two blades, dual jet engine turbine blades [also known as jet engine "modulation"] as well as multipath). The sampling rate of the helicopter data is 48,000 samples per second. The structure of the radar-illuminated Eurocopter BO-105, built in Germany, is illustrated in Fig. 1.2(c).

The fourth new signal provided for this guide is a not a time record. Rather, it is a grayscale image of 423 by 427 pixels in `texture.mat`. It will be used to compare 2-D spatial spectral analysis techniques in Chapter 5 of this Guide.

The next four chapters of this guide use demonstration scripts in MATLAB to illustrate and compare estimation performance of spectral techniques developed in *Digital Spectral Analysis*. Chapter 2 illustrates short duration signal analysis by techniques developed in Chapters 5–13 of *Digital Spectral Analysis*. Chapter 3 illustrates long duration signal analysis using the same techniques as used in Chapter 2, but modified for nonstationary longer duration signals. Chapter 4 illustrates multichannel signal analysis by techniques developed in Chapter 15 of *Digital Spectral Analysis*. Chapter 5 illustrates 2-D signal/image analysis by techniques developed in Chapter 16 of *Digital Spectral Analysis*.

Chapter 2

SHORT SIGNAL FREQUENCY ANALYSIS

The textbook *Digital Spectral Analysis* provides alternative spectral analysis techniques to classical Fourier-based frequency analysis based on the periodogram and correlation-based power spectral density (PSD) estimators. These alternative spectral techniques are especially useful for short duration signal sample records (typically 10s to 100s of samples). These alternative methods make additional assumptions about the signal that may yield better resolution and/or better estimates of the spectral features. To the degree that the alternative method assumptions approximate the measured signal frequency-domain features, estimation improvements can be expected. If the assumptions do not match (for example, additive white noise is expected but colored noise is the actual case or, as another example, that the assumed signal features are sinusoidal but there are no sinusoids in the signal), worse results may occur. The MATLAB demonstration scripts and the provided data will permit a user to see for themselves the behavior of each technique relative to other techniques. Each script will follow the five operation steps outlined in Chapter 1. Four of the five steps are common to all the scripts in this chapter. Only the algorithm and parameter selection step differs by method. All demonstration scripts will compute properly for either real-valued or complex-valued signal sample vectors (in column vector format). A logarithmic plotting axis is used for the spectral magnitude by computing $10\log_{10}(PSD(f)/maxPSD)$ (dB units) in which $maxPSD$ is the maximum PSD value across all frequencies as a reference. This means the maximum dB value will be 0. Logarithmic scaling is necessary to capture all the detail in the spectrum, including very weak signal components that often are of most interest.

Demonstration Script: `spectrum_demo_per`

This script implements the Welch-based method of classical Fourier-based periodogram spectral analysis. Figure 2.1 graphically illustrates the three input

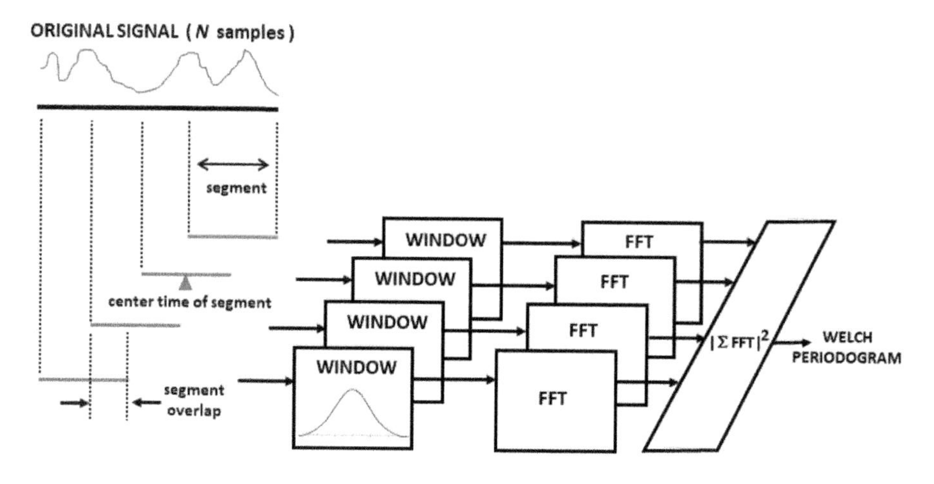

Figure 2.1: Graphic illustrating the three periodogram input estimation parameters.

parameters that must be selected for processing the signal record. If there is a need to reduce spectral estimation variance, a number of sample spectra must be averaged to achieve a reduction in statistical variance. The parameter **seg_size** will determine the sample segment size used for each sample spectrum. Overlap of adjacent segments should be done to use samples more than once. Selecting `seg_overlap` of 50 percent is suggested as typically it is the best statistical choice, although specific signal characteristics (for example, very fast frequency changes) may dictate more or less overlap. A window must be selected to reduce sidelobes due to the finite segment duration, but always at the expense of a slight reduction in spectral resolution. A full dialogue running `spectrum_demo_per` looks like this

```
This script demonstrates PERIODOGRAM spectral analysis.

** SOURCE **
Signal source choices:
    test1987              [1]
    doppler_radar         [2]
    bat_ultrasonic_pulse  [3]
    doppler_heart         [4]
    doppler_helicopter    [5]
Enter signal source number: 2

** SHORT SIGNAL vs LONG SIGNAL ANALYSIS **
Short signal analysis produces a 1-D frequency-only spectrum.
Long signal analysis produces a 2-D localized frequency vs localized time gram.
Select short signal [1] or long signal [2] analysis: 1
Selected analysis choice will use 64 signal samples, a duration of 0.0256 sec

** PERIODOGRAM SPECTRAL ESTIMATOR PARAMETERS **
Select window None [0], Hamming [1], or Nuttall [2]: 1
```

```
Enter segment size (# samples) per sample spectrum: 48
Enter overlap (# samples) between segment intervals: 24

** PERFORMING SPECTRAL ANALYSIS BASED ON SELECTIONS **

** PLOT PARAMETERS **
Range in dB to be plotted (suggest initially 60 dB): 50
Frequency axis scaling choices:
     actual   --  frequency expressed in Hertz
     fraction --  frequency expressed as fraction of sampling frequency
Enter string for scaling choice: actual
```

which produces the plotted spectral results in Fig. 2.2(b). For all demonstration scripts in this chapter, always select 1 for short signal. If one selects no window (0), segment size 64 (the entire available data record), and no overlap (0 since there is no additional data available to overlap if selecting the entire data record), the results are shown in Fig.2.2(a). These parameter selections should produce the highest resolution, yet the two close signals at 500 and 525 Hz are not resolved. There are two frequency axis scaling choices. The fraction of sampling frequency is a dimensionless frequency axis scaling in which $f_{actual}/f_{sample\_rate}$ is assumed. This scaling is sometimes useful when the sampling rate is not known

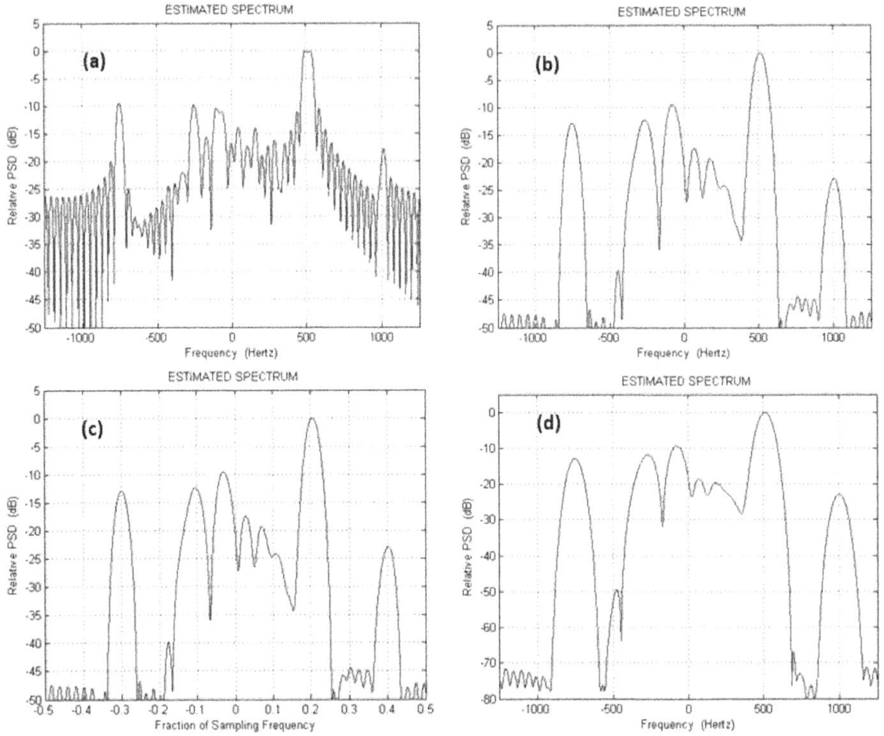

Figure 2.2: Four cases of periodogram estimate of data `doppler_radar`.

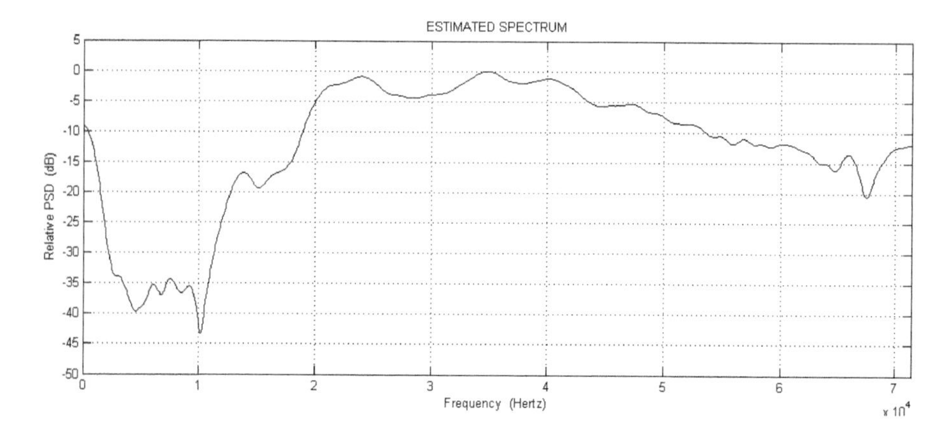

Figure 2.3: Periodogram analysis of bat echolocation pulse.

or is uncertain. If this is selected, complex signals will have a fractional frequency span from -0.5 to $+0.5$. If a real signal, the fractional frequency axis will automatically span 0 to $+0.5$ as the -0.5 to 0 range is a mirror image in the case of real-valued signals and therefore need not be plotted. If the script above is run, and only change plot parameter `actual` with `fraction`, the results are shown in Fig. 2.2(c). If the script above is run, and the window is changed to 2 (Nuttall) and the plotting range is changed to 80 dB rather than 50 dB, the results are shown in Fig. 2.2(d). This shows that the Nuttall window significantly reduces the sidelobes, but at the expense of further reducing the resolution (widened main lobe width).

If the selections that create the spectral plot are not satisfactory, just the last operational section of the script (plotting) can be run again without starting the demo script from the beginning. Simply execute `plot_psd` from the MATLAB command window and one may adjust the plotting parameters.

Another illustration of periodogram analysis is to select source 3 (bat pulse), window 1 (Hamming), segment size of 128 samples, and overlap of 64 samples, which results in the spectral estimate shown in Fig. 2.3. At first glance, it appears that the bat signal has a very broadband frequency span without much detail. This will be examined again with procedures covered in Chapter 2 that will bring out the frequency features not seen in short duration signal analysis.

Demonstration Script: `spectrum_demo_corr`

The other classical Fourier-based estimator generates the spectral estimate by first estimating a finite autocorrelation sequence (ACS) from the data record. Using script `plot_correlation`, either the unbiased or the biased autocorrelation estimates for data `doppler_radar` may be plotted, as illustrated in Fig. 2.4 for lag 0 up to the maximum lag 63 (one less than the data record sample length).

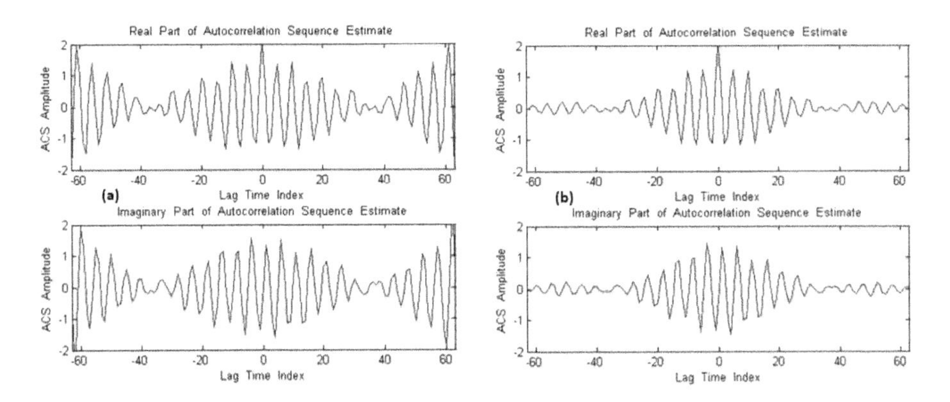

Figure 2.4: Complex-valued autocorrelation estimate of doppler radar data. (a) Unbiased. (b) Biased.

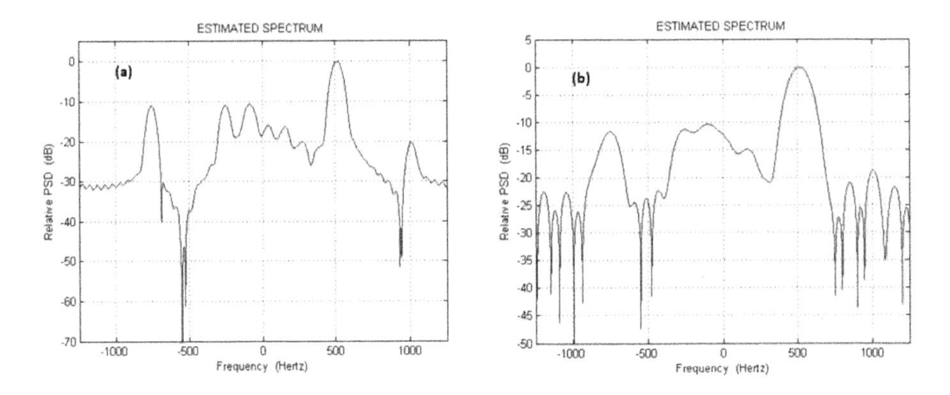

Figure 2.5: Correlation-based spectral estimate of data `doppler_radar`.

Note that biased estimate is the unbiased with a triangular window applied to de-emphasize the ACS higher lag indices due to their greater statistical estimation variance. The script for this technique is based on the pioneering Blackman and Tukey approach from the 1950s (favorite classical method before PC availability and the FFT algorithm of Fourier analysis introduced in 1965). Unbiased ACS is estimated inside the script, to which one of three window choices can be applied. The script is identical to script `spectrum_demo_per` except for the technique parameter inputs, which are

```
** BLACKMAN-TUKEY CORRELATION-BASED SPECTRAL ESTIMATOR PARAMETERS **
Select window  None [0], Hamming [1], or Nuttall [2]: 1
Enter maximum lag (# samples) for estimated ACS (and CCS if cross): 16
```

Applying the above parameter choices to data choice 3, the plotted results are shown in Fig. 2.5(b). The choice of maximum lag 16 is selected to reflect the same degrees of freedom given to alternative spectral estimators for which 16 is an appropriate parameter choice for those estimators. This permits an examination across all spectral estimators for quality of spectral estimate using the same degrees of freedom for all methods. Fig. 2.5(a) shows the spectral estimate results for window choice 2 and maximum lag 48, which allows one to compare with Fig. 2.2(b) that also uses a segment size 48.

Demonstration Script: spectrum_demo_ar

This script illustrates autoregressive spectral analysis techniques. There are five AR parameter estimation algorithm choices that were developed in text *Digital Spectral Analysis*. The AR demonstration script produces an identical dialogue as spectrum_demo_per except for the AR specific algorithm and parameter operating section, which is

```
 ** AUTOREGRESSIVE SPECTRAL ESTIMATOR PARAMETERS **
 Autoregressive (AR) spectral analysis algorithm choices:
      1  --  Yule-Walker
      2  --  Lattice (Geometric)
      3  --  Lattice (Burg)
      4  --  Covariance Linear Prediction
      5  --  Modified Covariance Linear Prediction
Enter number of AR algorithm choice: 3
Enter autoregressive spectral order (AR filter length) to use: 16
```

Two estimation algorithms are illustrated here. The Burg algorithm is arguably the most well known of the five techniques. Fig. 2.6(a) shows the resulting AR PSD plot for the selection of the script parameters above. Note that a plot dB range of 80 was necessary to capture detail in the spectral estimate. AR spectra have no sidelobes unlike the classical methods, and therefore a greater dB plotting range is required to highlight both strong and weak signal components.

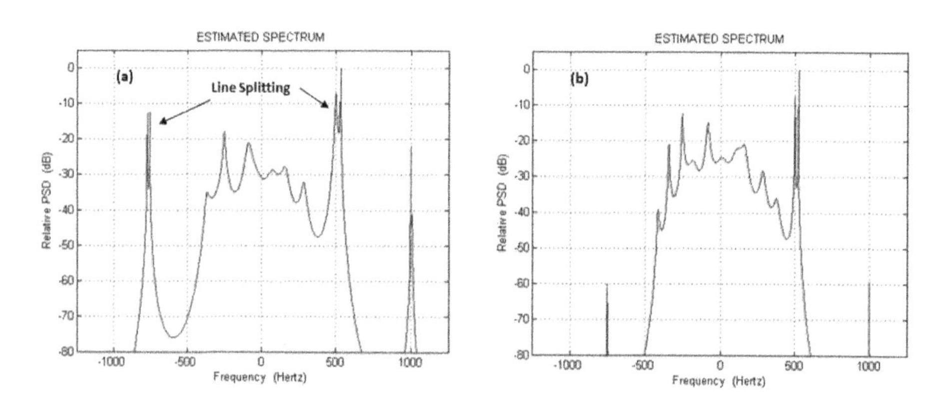

Figure 2.6: Autoregressive AR(16) spectral estimate of data doppler_radar. (a) Burg lattice algorithm. (b) Modified covariance algorithm.

If the algorithm choice 5 (modified covariance) and order 16 are used, the results are shown in Fig. 2.6(b). Note that the Burg algorithm exhibits spectral line splitting (two close sharp peaks are estimated where there should only be one), whereas the modified covariance algorithm correctly shows single sharp peaks.

Demonstration Script: `spectrum_demo_rls`

This script illustrates time-recursive AR spectral estimation. There are two time-recursive algorithms from which to choose (LMS or RLS). The convergence rate to accurate AR parameter estimates vary from 100s to 1,000s of samples, and the frequency content of the signal must remain unchanging or very slowly changing during convergence so that the AR parameter estimation can slowly follow the changes. None of the signal sources provided for this user guide have these characteristics, so only the use of the source code can be illustrated and no data record is provided to illustrate convergence performance. If the script selections

```
** TIME RECURSIVE AR SPECTRAL ESTIMATOR PARAMETERS **
Time recursive AR spectral analysis algorithm choices:
     1  --  Least Mean Square (LMS)
     2  --  Fast Recursive Least Squares (RLS)
Enter number of AR algorithm choice: 1
Enter autoregressive spectral order (AR filter length) to use: 16
Enter adaptive factor value (mu for lms  or  omega for fast_rls): 2.e-2
```

are made, the resulting AR spectral estimate is shown in Fig. 2.7(a).

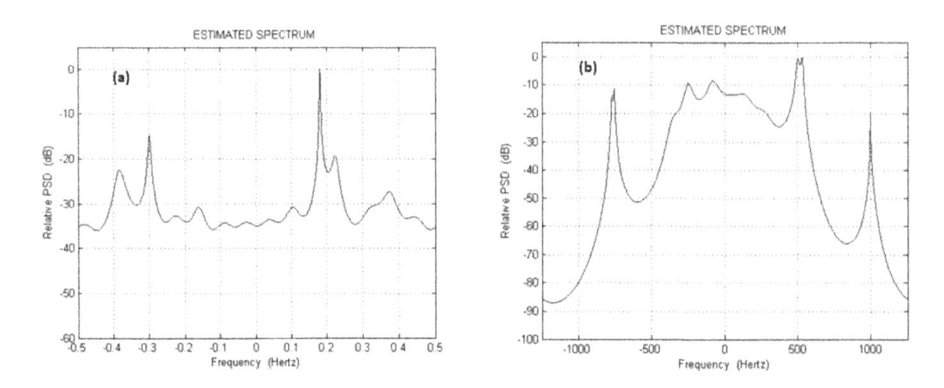

Figure 2.7: (a) Time recursive AR spectral analysis by LMS algorithm. (b) Minimum variance spectral analysis.

Demonstration Script: spectrum_demo_minvar

This script illustrates minimum variance spectral estimation. The section of the script related to minimum variance spectral estimation has only one parameter:

```
** MINIMUM VARIANCE SPECTRAL ESTIMATOR PARAMETER **
Enter order of minimum variance spectral filter (#taps=order+1): 16
```

The results using the doppler radar data are shown in Fig. 2.7(b). Note that a plotting range of 80 dB is needed to capture all detail. Of all the alternative spectral estimators to classical spectral analysis, only the minimum variance spectral technique plot has approximately a linear relationship between the relative peak heights and the relative strengths of the signals.

Demonstration Script: spectrum_demo_ma

This script illustrates moving average (MA) spectral estimation. The parameter selection portion of the script asks only for a single order parameter.

```
** MOVING AVERAGE SPECTRAL ESTIMATOR PARAMETERS **
Enter moving average spectral order (MA filter length) to use: 16
```

The results are shown in Fig. 2.8(a).

Demonstration Script: spectrum_demo_arma

This script illustrates autoregressive moving average (ARMA) spectral estimation. The parameter selection portion of the script asks for only the two order parameters.

```
** ARMA SPECTRAL ESTIMATOR PARAMETERS **
Enter autoregressive spectral order (AR filter length) to use: 16
Enter moving average spectral order (MA filter length) to use: 16
```

The results are shown in Fig. 2.8(b) usinga plot dB range of 80.

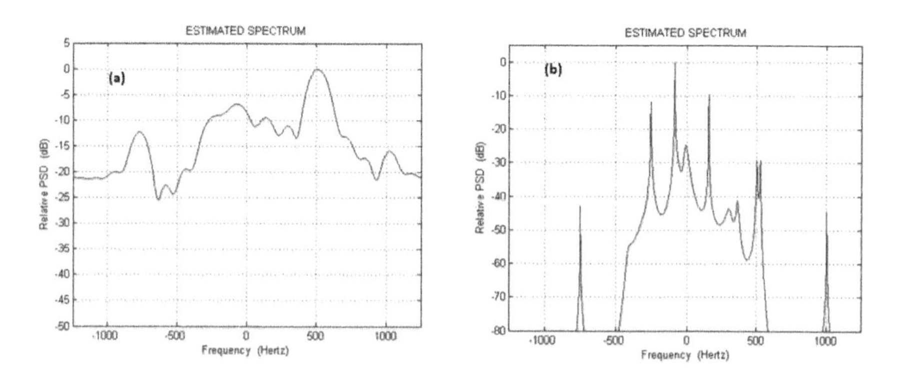

Figure 2.8: (a) MA(16) spectral estimate. (b) ARMA(16,16) spectral estimate.

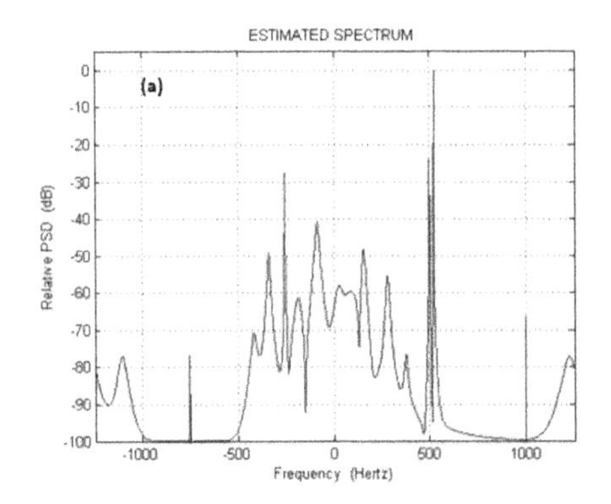

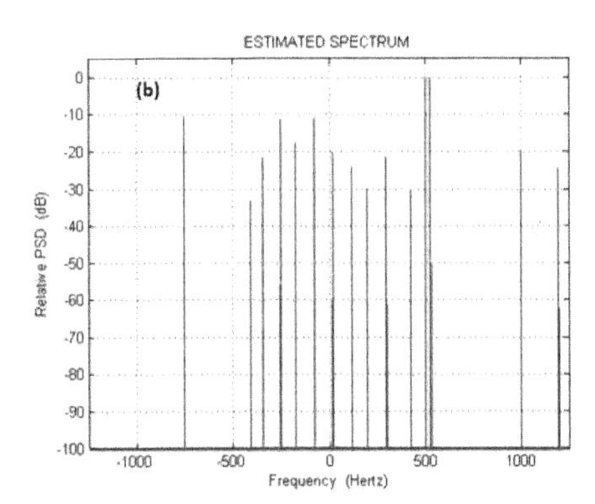

Figure 2.9: Prony estimated spectra. (a) Damped sinusoidal model. (b) Undamped sinusoidal model.

Demonstration Script: `spectrum_demo_prony`

This script demonstrates the Prony approach for both exponential modeling and a derived energy spectral density estimate. Two algorithm techniques are coded. One is the normal Prony approach that places no constraints on the exponential model parameter estimates, which creates in general a sum of damped sinusoidals model. The other is a modified Prony approach that places a constraint on the exponential model parameter estimation, yielding a sum of undamped sinusoidals model. In the parameter selection operational section of the script, if the following is entered:

```
** PRONY EXPONENTIAL MODEL & SPECTRAL ESTIMATOR PARAMETERS **
Prony exponential estimation algorithm choices:
     1  --  normal least squares Prony method (damped exponentials)
     2  --  modified Prony method (undamped exponentials)
Enter number of Prony algorithm choice: 1
Enter number of exponential components to be estimated: 16

** PERFORMING SPECTRAL ANALYSIS BASED ON SELECTIONS **

List last-computed exponential parameters? 1--Yes, 2--No : 2
```

then the results shown on Fig. 2.9(a) is produced. To illustrate the modified Prony, the algorithm choice entered is 2, and the post processing question is answered 1 for yes, which produces both a parameter estimation listing

```
 ** PERFORMING SPECTRAL ANALYSIS BASED ON SELECTIONS **
List last-computed exponential parameters? 1--Yes, 2--No : 1

Estimated Real-value Exponential Parameters
    Frequency (Hz)    Damping (1/sec)       Amplitude           Phase (rad)
  1.0e+03 times
```

```
-0.749999541368881* -0.000000000000001   0.000310047070885  -0.001895925466302
-0.405649335246189  -0.000000000000039   0.000022832127660  -0.001599774776026
-0.339984110039089   0.000000000000219   0.000086319950795  -0.001060702746595
-0.251140178171685* -0.000000000000730   0.000280769060897  -0.000458666008677
-0.175895047568430   0.000000000000978   0.000138788232242  -0.003024243856928
-0.078500893668000  -0.000000000000664   0.000285081596857  -0.000366487194373
 0.019539564627203   0.000000000000904   0.000100981766633  -0.000019676738108
 0.118349122039657  -0.000000000002052   0.000064531124246  -0.002184786957338
 0.201950674586705   0.000000000002304   0.000032438074564   0.002446273426219
 0.300352036004069  -0.000000000001159   0.000086019702751   0.001943126853104
 0.424983827852483   0.000000000000424   0.000032269871392  -0.000213046390200
 0.500550341982215* -0.000000000000271   0.001033863197798   0.001208272556224
 0.524933581641777*  0.000000000000108   0.001015852191964   0.001368372744900
 0.999996667223860*  0.000000000000008   0.000103581428481   0.002585236161679
 1.198292242728012   1.182921497697304   0.000000000000000   0.002219359491132
 1.198292242728012  -1.182921497697318   0.000061533397196  -0.002186869843777
```

and the plot results of Fig. 2.9(b). Note that the power of the Prony approach is in the accuracy of its parameter estimates. As indicated by * in the listing above for the five complex sinusoidals in the data, the five sinusoidal frequencies are estimated accurately to at least three decimal places even in the presence of colored noise (simulated clutter).

Demonstration Script: `spectrum_demo_eigen`

This script illustrates frequency estimation results by four eigenanalysis-based techniques (none are true spectral estimators as only sinusoidal frequencies and the variance of white noise is estimated). If the algorithm and parameter selection portion of the script enters these values:

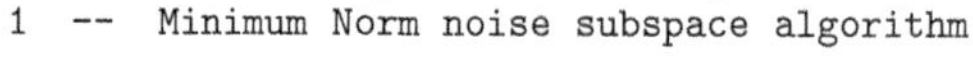

```
** EIGENANALYSIS ESTIMATOR PARAMETERS **
Eigenanalysis frequency estimation algorithm choices:
     1  --  Minimum Norm noise subspace algorithm
```

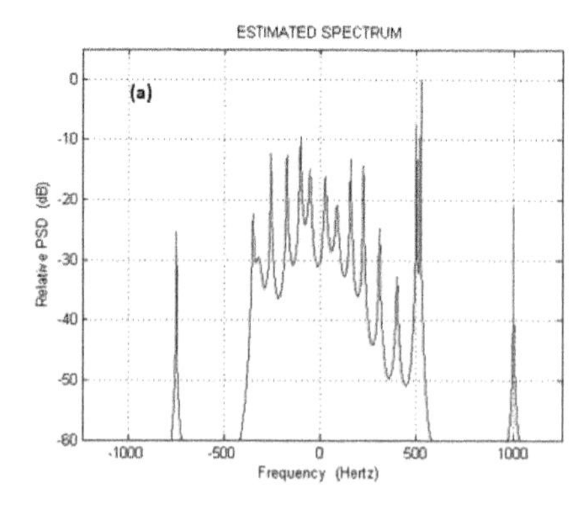

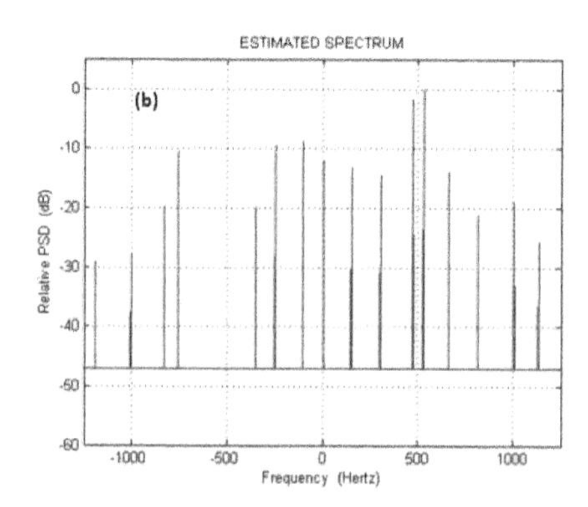

Figure 2.10: Frequency estimators by eigenanalysis. (a) MUSIC method. (b) Pisarenko method.

```
     2  --  MUSIC (MUltiple SIgnal Classification) noise subspace
     3  --  EV (EigenVector) noise subspace algorithm
     4  --  Pisarenko harmonic decomposition via noise subspace
 Enter number of eigenanalysis algorithm choice: 2
 Enter presumed number of complex sinusoids (2 for each real sine): 16
```

then the frequency estimation (peak locations) results are shown in Fig. 2.10(a). If the algorithm choice is selected as 4 (Pisarenko) the results are shown in Fig. 2.10(b).

Chapter 3

LONG SIGNAL TIME-FREQUENCY ANALYSIS

This chapter considers *long* signals, that is, signals that have 100s to 1,000s of samples which have highly time-varying frequency content (nonstationary). Although this was not covered in *Digital Spectral Analysis*, it is possible to use all the demo scripts in Chapter 2 to capture the time-varying behavior. To capture the time-varying detail, it is necessary to use a *short* sliding analysis interval stepped through the data, performing a spectral analysis at each step of just the data within the analysis interval. The multiple stepped analyses are "stacked" to form a localized time vs localized frequency representation of the time-varying spectral characteristics of the signal. Figure 3.1 illustrates the parameters of this approach. An analysis interval is selected and the number

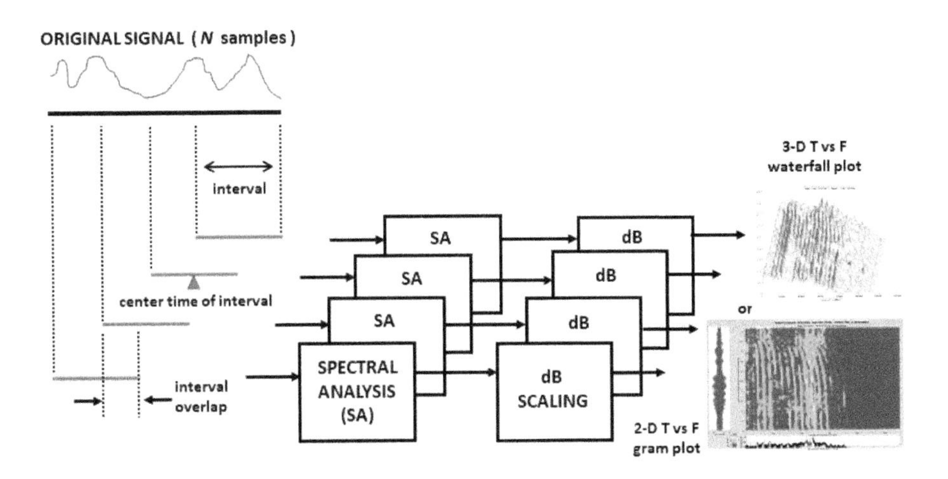

Figure 3.1: Graphical description of sliding analysis parameters for time-frequency analysis.

of samples overlap between adjacent analysis windows selected. Each analysis window is processed by any selected classical or alternative spectral estimator demonstrated in Chapter 2 of this user guide. In fact, the demonstration scripts already have incorporated long signal time-frequency analysis into the script. Please note that the sliding analysis "interval" of time vs frequency analysis, which applies to all spectral techniques, should not be confused with the "segment" selection of the periodogram spectral analysis that is used to average segment periodograms for purposes of variance reduction.

A 2-D time vs frequency array is created by performing long signal analysis. There are two plotting approaches provided to visualize the results. A "waterfall" type of plot layers the individual spectra in a 3-D presentation, as shown at the top right of Fig. 3.1. A "color gram" type of plot uses a flat color-coded 2-D image at the bottom right of Fig. 3.1. The "waterfall" plot can take a long computation time by MATLAB if the product of number of time points and the number of frequency points is greater than 5,000, so the "color gram" is the preferred plot as it is formed much quicker.

Performing long signal time vs frequency requires much experimentation with variations of interval size and overlap selections, along with acceptable plot settings to capture low level signal behavior.

Demo: Bat Ultrasonic Pulse Time-Frequency Analysis

In Chapter 1, the bat ultrasonic pulse was thought to not have significant spectral features when analyzed under short signal analysis procedures (see Fig. 2.3). However, performing a long signal analysis (even though the signal is only 400 samples) under the following script inputs for a periodogram analysis

```
This script demonstrates PERIODOGRAM spectral analysis.

** SOURCE **
Signal source choices:
    test1987              [1]
    doppler_radar         [2]
    bat_ultrasonic_pulse  [3]
    doppler_heart         [4]
    doppler_helicopter    [5]
Enter signal source number: 3

** SHORT SIGNAL vs LONG SIGNAL ANALYSIS **
Short signal analysis produces a 1-D frequency-only spectrum.
Long signal analysis produces a 2-D localized frequency vs localized time gram.
Select short signal [1] or long signal [2] analysis: 2
Enter analysis interval duration (# samples): 40
Enter overlap (# samples) between analysis intervals: 39
Selected analysis choice will use 400 signal samples, a duration of
  0.0028 seconds.

** PERIODOGRAM SPECTRAL ESTIMATOR PARAMETERS **
```

```
Select window None [0], Hamming [1], or Nuttall [2]: 1
Enter segment size (# samples) per sample spectrum. Suggest selecting
the same as the analysis interval duration (40): 40
Enter overlap (# samples) between segment intervals.
Suggest setting segment overlap = 0: 0

** PERFORMING SPECTRAL ANALYSIS BASED ON SELECTIONS **

** PLOT PARAMETERS **
The analysis parameters yielded 361 gram lines to plot.
Enter text for time-frequency gram title: BAT ULTRASONIC SIGNAL SOURCE
Enter 3 [3-D waterfall] or 2 [2-D color image] to display time-frequency gram: 2
PSD gram of time-frequency analysis will be scaled and plotted in
  logarithmic units (dB).
Largest PSD value will be reference at 0 dB. Smaller PSD values will be
  negative dB.
dB top and dB bottom will now be specified. dB values above dB_top will be set
  to dB_top.
dB values below dB_bottom will be set to dB_bottom.
Enter dB_top choice: -10
Enter dB_bottom choice: -40
Time vs frequency axes scaling choices:
     0 --  samples vs fraction of sampling frequency [dimensionless frequency
           axis]
     1 --  seconds (s) vs  Hertz (Hz) [suggest for test1987, doppler_radar, and
           heart data]
     2 --  milliseconds (ms) vs KiloHertz (KHz) [suggest for bat and helicopter
           data]
Enter axes scaling choice: 2
Enter spacing in KHz between frequency axis ticks: 10
Enter spacing in milliseconds between time axis ticks: .2
Waveform is real-valued. Plot left of gram.
```

showing that the bat pulse frequency response has several features that vary rapidly with time (see color gram in Fig. 3.2). Note that the real-valued time waveform is plotted adjacent to the left of the gram. A red line appears at the bottom of the waveform plot to show the extent of the sliding analysis interval. The magnitude of an FFT of the entire waveform is plotted at the bottom of the gram. A color bar of the dB coding levels used in the time-frequency gram may be found at the bottom left corner of the figure. The selection of db_top $= -10$ and dB_bottom $= -40$ means that any spectral values between 0 dB and -10 dB are set to red (top color in color bar) and values -40 dB and below are set to dark violet (bottom color of color bar).

Rerunning the script with parameter change `interval overlap = 20` from `39` and selecting the waterfall plot choice, and providing the following plot parameter responses:

```
** PLOT PARAMETERS **
The analysis parameters yielded 19 gram lines to plot.
Enter text for time-frequency gram title: BAT ULTRASONIC PULSE
Enter 3 [3-D waterfall] or 2 [2-D color image] to display time-frequency gram: 3
Time vs frequency axes scaling choices:
   0 --  samples vs fraction of sampling frequency [dimensionless frequency axis]
```

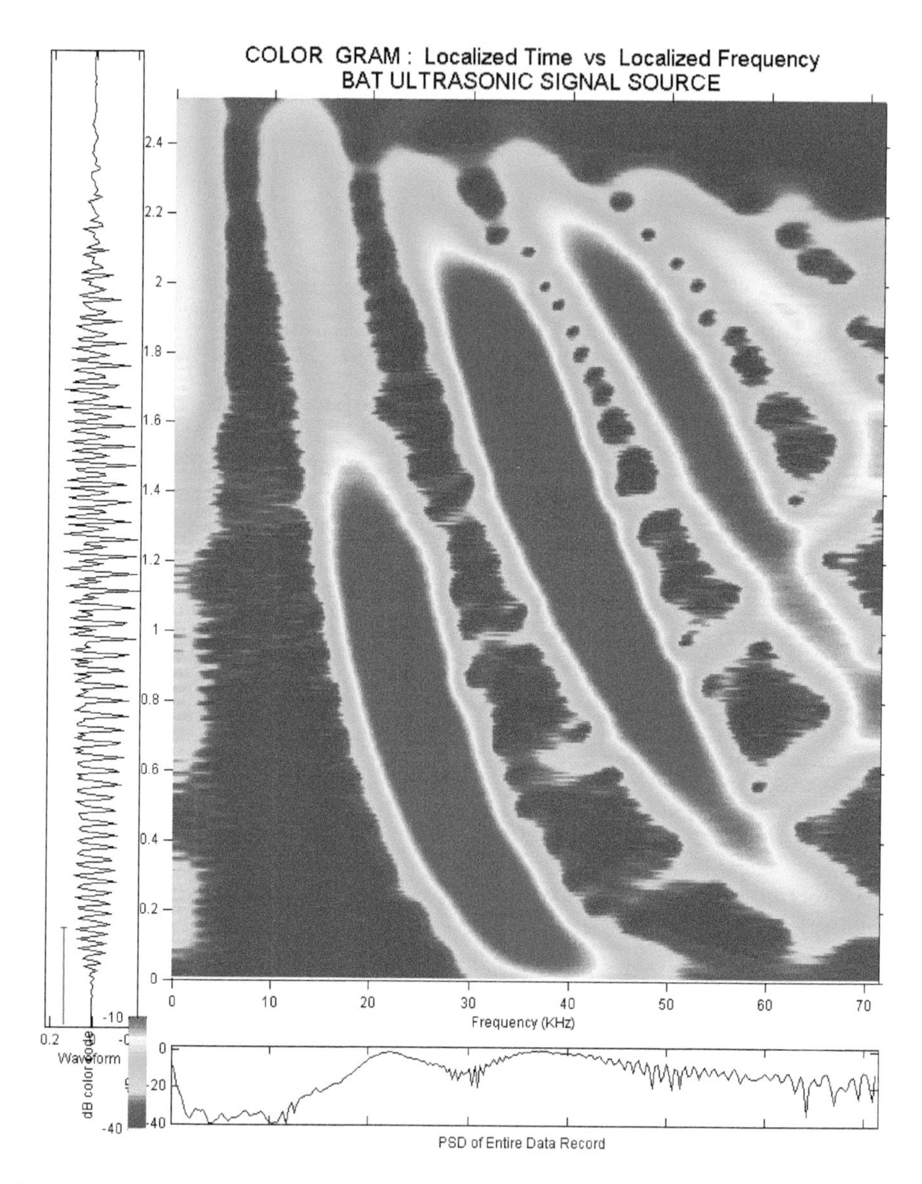

Figure 3.2: Bat ultrasonic pulse long signal periodogram analysis using gram plot.

```
   1 --  seconds (s) vs  Hertz (Hz) [suggest for test1987, radar, heart data
   2 --  milliseconds vs KHz [suggest for bat and helicopter data]
Enter axes scaling choice: 2
Data is real-valued.
PSD gram of time-frequency analysis will be scaled in logarithmic units (dB).
Logarithmic range in dB that PSD is to be plotted: 50
```

will yield the plot shown in Fig. 3.3.

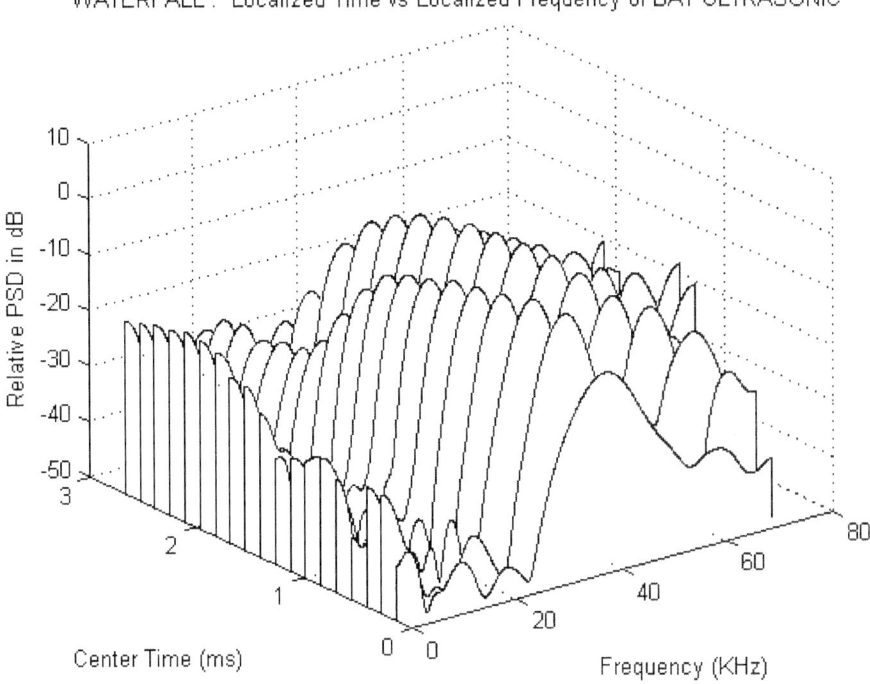

Figure 3.3: Bat ultrasonic pulse long signal periodogram analysis using waterfall plot.

The selection of the "optimal" analysis window of $N = 40$ samples to reveal signal frequency behavior with minimal smearing of detail was determined experimentally. However, there are methods to analytically determine an analysis sample size. As many features in the bat signal are nearly linear, consider this approach for determining N. It was shown in *Digital Spectral Analysis* that the Fourier transform of a windowed data interval of duration NT_s sec, in which T_s is the sampling interval, has a frequency resolution of approximately $1/(NT_s)$ Hz bandwidth. Thus, a criterion for having a roughly stationary statistical behavior within an analysis window is any change in frequency content (possibly causing a plot smear) between analysis intervals must be less than the frequency resolution of the analysis (since otherwise it would not be resolvable), in which case the signal is considered to be essentially stationary over the interval of N samples.

Analysis Duration $*$ Frequency Rate of Change $\leq$ Spectrum Resolution

or

$$(NT_s)(\Delta F/\Delta T) \leq (1/NT_s)$$

yielding

$$N^2 \leq (2/T_s^2)(\Delta T/\Delta F)$$

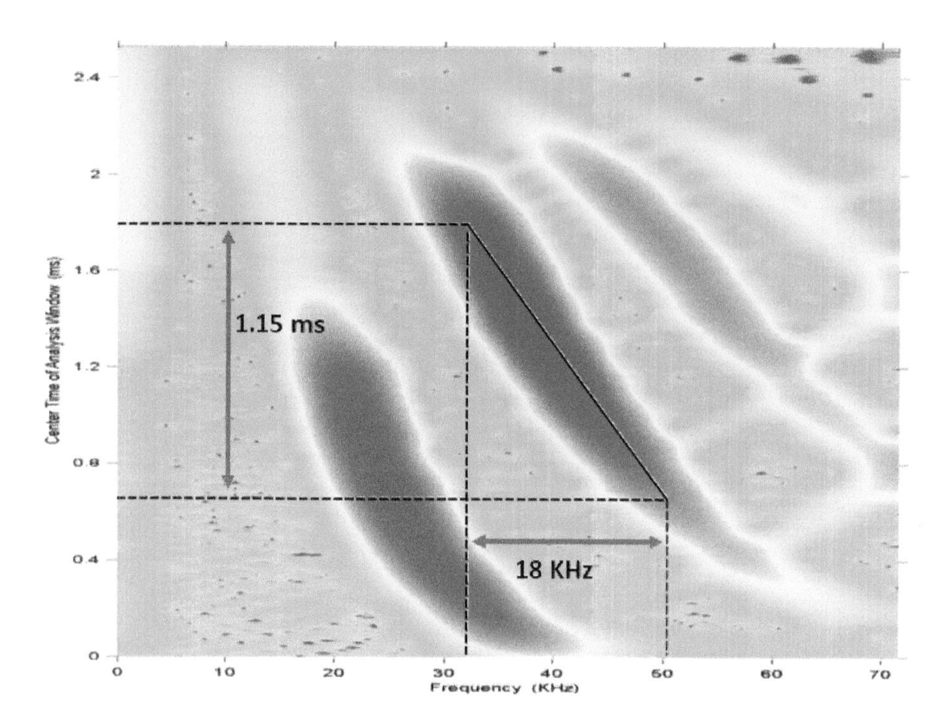

Figure 3.4: Technique to estimate analysis interval size using bat signal.

Using the bat time vs frequency gram shown in Fig. 3.4, the nearly linear feature marked on this figure has lines that indicate frequency change and time change to calculate the slope of rate of change. Substituting the numbers from Fig. 3.4 and noting that $T_s = 7$ microseconds, the formula above yields $N \leq 36.11$ samples. This compares favorably with the experimentally determined 40 sample interval that appears to yield the sharpest overall color gram.

Demo: Heart Ultrasound Time-Frequency Analysis

Next, an illustration is provided for the frequency (blood flow velocity) time evolution of blood flow rate within the heart from an ultrasonic probe sample sequence measured at a depth of approximately 9 cm and located at a heart valve. An alternative spectral analysis will be compared to classical FFT-based periodogram analysis. The script `spectrum_demo_per` is processed with the following inputs

```
This script demonstrates PERIODOGRAM spectral analysis.

** SOURCE **
Signal source choices:
    test1987            [1]
```

```
    doppler_radar        [2]
    bat_ultrasonic_pulse [3]
    doppler_heart        [4]
    doppler_helicopter   [5]
Enter signal source number: 4

** SHORT SIGNAL vs LONG SIGNAL ANALYSIS **
Short signal analysis produces a 1-D frequency-only spectrum.
Long signal analysis produces a 2-D localized frequency vs localized time gram.
Select short signal [1] or long signal [2] analysis: 2
Enter analysis interval duration (# samples): 32
Enter overlap (# samples) between analysis intervals: 24
Selected analysis choice will use 16384 signal samples, duration 2.6214 seconds.

** PERIODOGRAM SPECTRAL ESTIMATOR PARAMETERS **
Select window None [0], Hamming [1], or Nuttall [2]: 1
Enter segment size (# samples) per sample spectrum. Suggest selecting
the same as the analysis interval duration (32): 32
Enter overlap (# samples) between segment intervals.
Suggest setting segment overlap = 0: 0

** PERFORMING SPECTRAL ANALYSIS BASED ON SELECTIONS **

** PLOT PARAMETERS **
The analysis parameters yielded 2045 gram lines to plot.
Enter text for time-frequency gram title: ULTRASOUND DOPPLER OF HEART
Enter 3 [3-D waterfall] or 2 [2-D color image] to display time-frequency gram: 2
PSD gram of time-frequency analysis will be scaled and in logarithmic units (dB).
Largest PSD value will be reference at 0 dB. Smaller PSD values are negative dB.
dB top and dB bottom will now be specified. dB values above dB_top set to dB_top.
dB values below dB_bottom will be set to dB_bottom.
Enter dB_top choice: -10
Enter dB_bottom choice: -60
Time vs frequency axes scaling choices:
   0 --  samples vs fraction of sampling frequency [dimensionless frequency axis]
   1 --  seconds (s) vs  Hertz (Hz) [suggest test1987, doppler_radar, heart data]
   2 --  milliseconds (ms) vs KiloHertz (KHz) [suggest bat and helicopter data]
Enter axes scaling choice: 2
Enter spacing in KHz between frequency axis ticks: 1
Enter spacing in milliseconds between time axis ticks: 500
Waveform is complex-valued. Real part left & imag part right of gram.
```

The results are shown in Fig. 3.5 for nearly three beat cycles. One of the interesting features is the evidence of aliasing on the left side of the color gram. Based on the diagram of Fig. 3.6, it is a simple task to move a portion of the aliased spectrum to the opposite side to rejoin aliased and unaliased spectra to improve diagnostic utility. One may wonder why the sample rate of 6,250 samples per second was not raised to prevent aliasing. There is a physical limitation in medical ultrasound that prevents raising the sample rate. As a propagating pulse waveform, an ultrasonic wave nominally travels through body tissue nominally at 1,540 meter/sec rate. One cannot launch another pulse until the previous pulse round trip time through the human body is complete, otherwise the pulse responses overlap and cannot be separated. For the probe location that acquires the signal record provided in this user guide, that limitation was

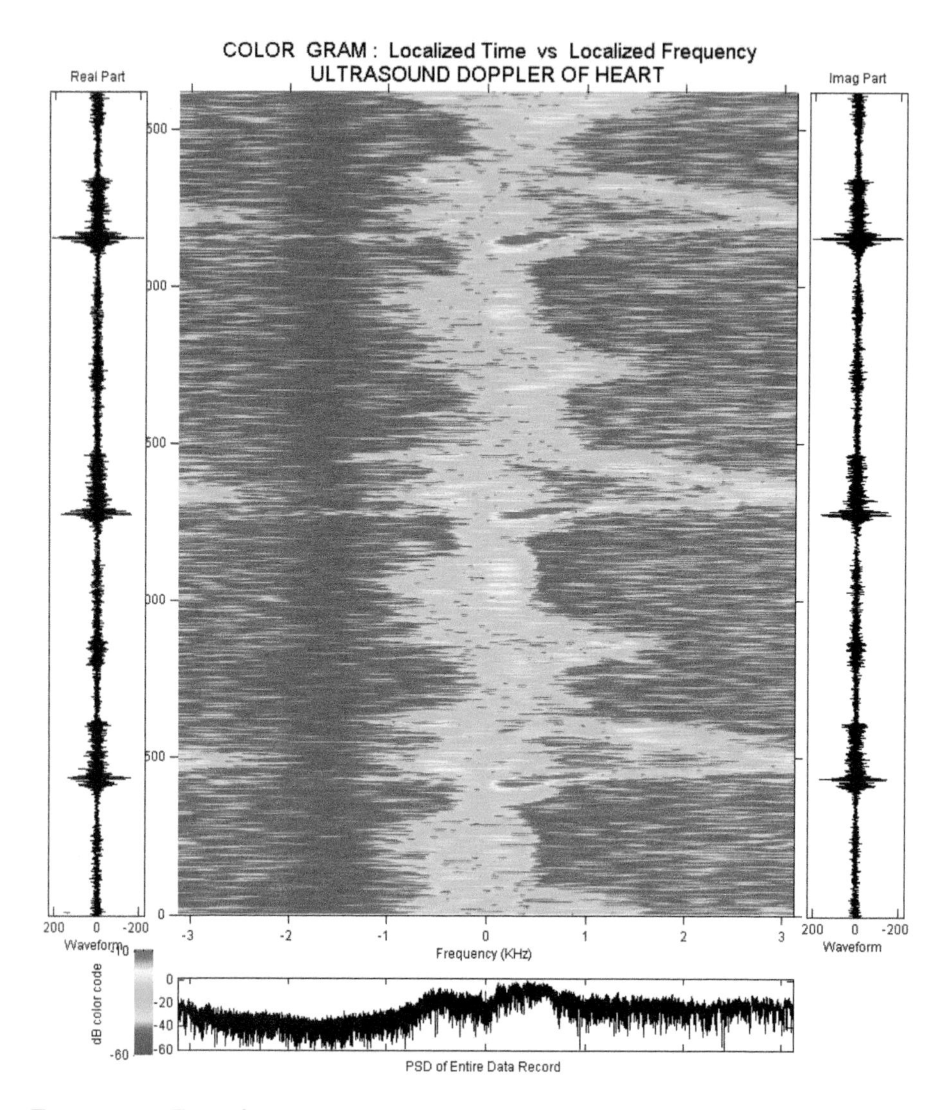

Figure 3.5: Time-frequency gram of cardiac blood flow process doppler by periodogram analysis.

6,250 pulses per second. Aliasing isolated in localized time-frequency space can be recovered as illustrated in Fig. 3.6.

An improvement in the detail of the blood flow components can be achieved by using autoregressive (AR) spectral analysis. The script `spectrum_demo_ar` is run with identical parameters as the periodogram script, except the modified covariance algorithm (5) is selected for order 16 and dB top and bottom boundaries of -40 to -80 dB. The results are shown in Fig. 3.7. Note the

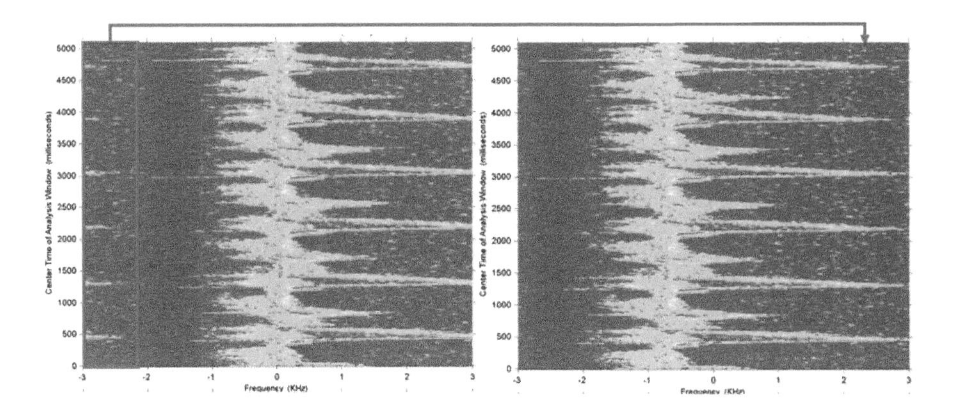

Figure 3.6: Frequency band shifting portions of the time-frequency gram to correct for aliasing.

sharper delineations in various doppler tracks that were not discernable in the periodogram color gram.

Demo: In-Flight Helicopter Radar Doppler Time-Frequency Analysis

As a final demonstration comparing a highly nonstationary signal record processed by periodogram analysis versus alternative minimum variance spectral analysis, this section will use the doppler radar signal of a helicopter in flight. The script `spectrum_demo_per` is run with the following parameter choices

```
This script demonstrates PERIODOGRAM spectral analysis.

** SOURCE **
Signal source choices:
    test1987             [1]
    doppler_radar        [2]
    bat_ultrasonic_pulse [3]
    doppler_heart        [4]
    doppler_helicopter   [5]
Enter signal source number: 5

** SHORT SIGNAL vs LONG SIGNAL ANALYSIS **
Short signal analysis produces a 1-D frequency-only spectrum.
Long signal analysis produces a 2-D localized frequency vs localized time gram.
Select short signal [1] or long signal [2] analysis: 2
Enter analysis interval duration (# samples): 48
Enter overlap (# samples) between analysis intervals: 40
Selected analysis choice will use 4096 signal samples, a duration of
  0.085333 seconds.

** PERIODOGRAM SPECTRAL ESTIMATOR PARAMETERS **
Select window None [0], Hamming [1], or Nuttall [2]: 1
```

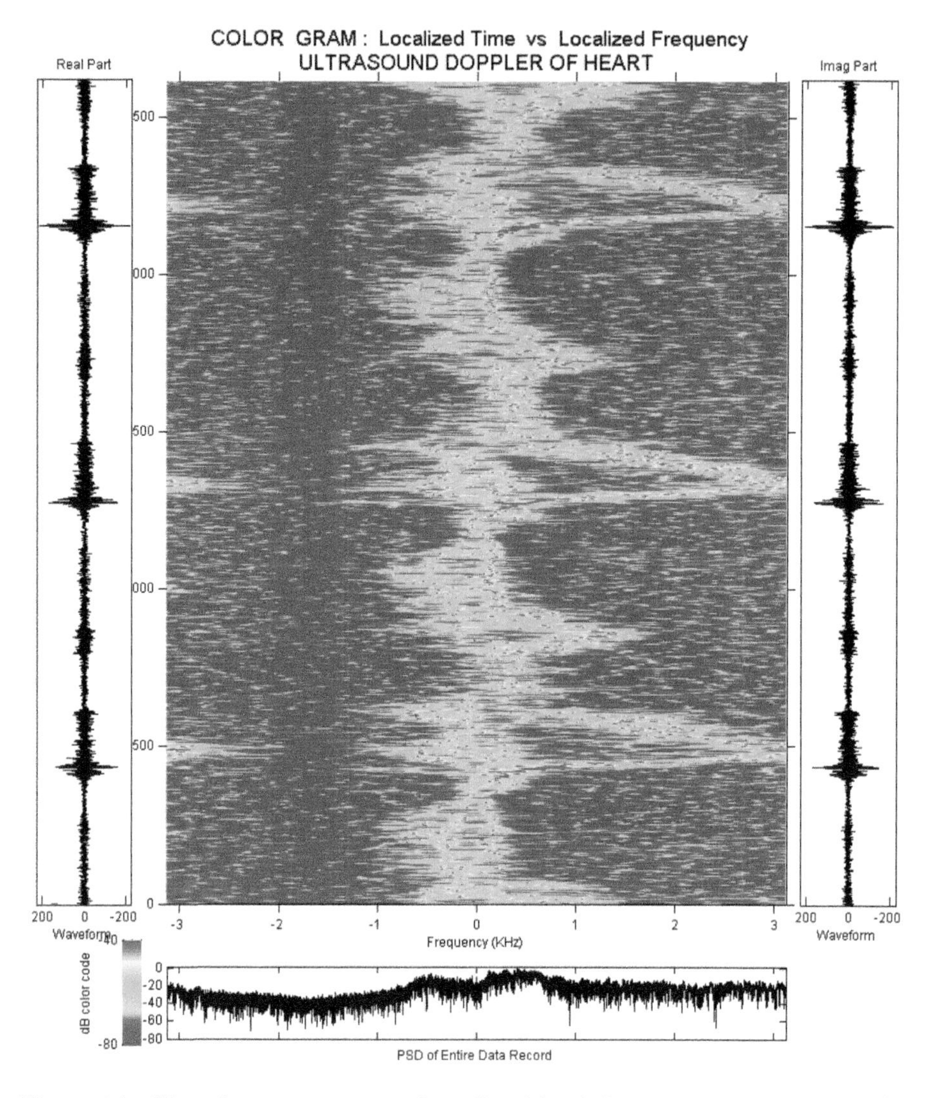

Figure 3.7: Time-frequency gram of cardiac blood flow process doppler by AR modified covariance algorithm.

```
Enter segment size (# samples) per sample spectrum. Suggest selecting
the same as the analysis interval duration (48): 48
Enter overlap (# samples) between segment intervals.
Suggest setting segment overlap = 0: 0

** PERFORMING SPECTRAL ANALYSIS BASED ON SELECTIONS **

** PLOT PARAMETERS **
The analysis parameters yielded 507 gram lines to plot.
```

```
Enter text for time-frequency gram title (no quotes around text):
  RADAR DOPPLER OF HELICOPTER
Enter 3 [3-D waterfall] or 2 [2-D color image] to display time-frequency gram: 2
PSD gram of time-frequency analysis will be scaled and plotted in logarithmic
  units (dB).
Largest PSD value will be reference at 0 dB. Smaller PSD values will be
  negative dB.
dB top and dB bottom will now be specified. dB values above dB_top will be set to
  dB_top.
dB values below dB_bottom will be set to dB_bottom.
Enter dB_top choice: -15
Enter dB_bottom choice: -60
Time vs frequency axes scaling choices:
   0 --  samples vs fraction of sampling frequency [dimensionless frequency axis]
   1 --  seconds (s) vs  Hertz (Hz) [suggest for test1987, doppler_radar, and
         heart data]
   2 --  milliseconds (ms) vs KiloHertz (KHz) [suggest for bat and helicopter
         data]
Enter axes scaling choice: 2
Enter spacing in KHz between frequency axis ticks: 2
Enter spacing in milliseconds between time axis ticks: 50
Waveform is complex-valued. Real part left & imag part right of gram.
```

producing the results shown in Fig. 3.8. Sinusoidal patterns for the four main rotor blades are obvious, as well as periodic broad frequency flashes of the tail rotor. A wide response centered on 0 Hz is due to a time code placed on the recording tape; it is not a helicopter doppler feature.

The script `spectrum_demo_minvar` is then run with the same parameter selections except for the selection of order = 30 as the minimum variance spectral estimator parameter. It produces the time-frequency gram shown in Fig. 3.9 for dB thresholds −10 to −60 dB. The spectrum is more resolved. The smear around the strong 3 KHz fuselage doppler in Fig. 3.8 is now resolved to where small micro-doppler variations can be seen, possibly evidence of rotating features on the main rotor hub.

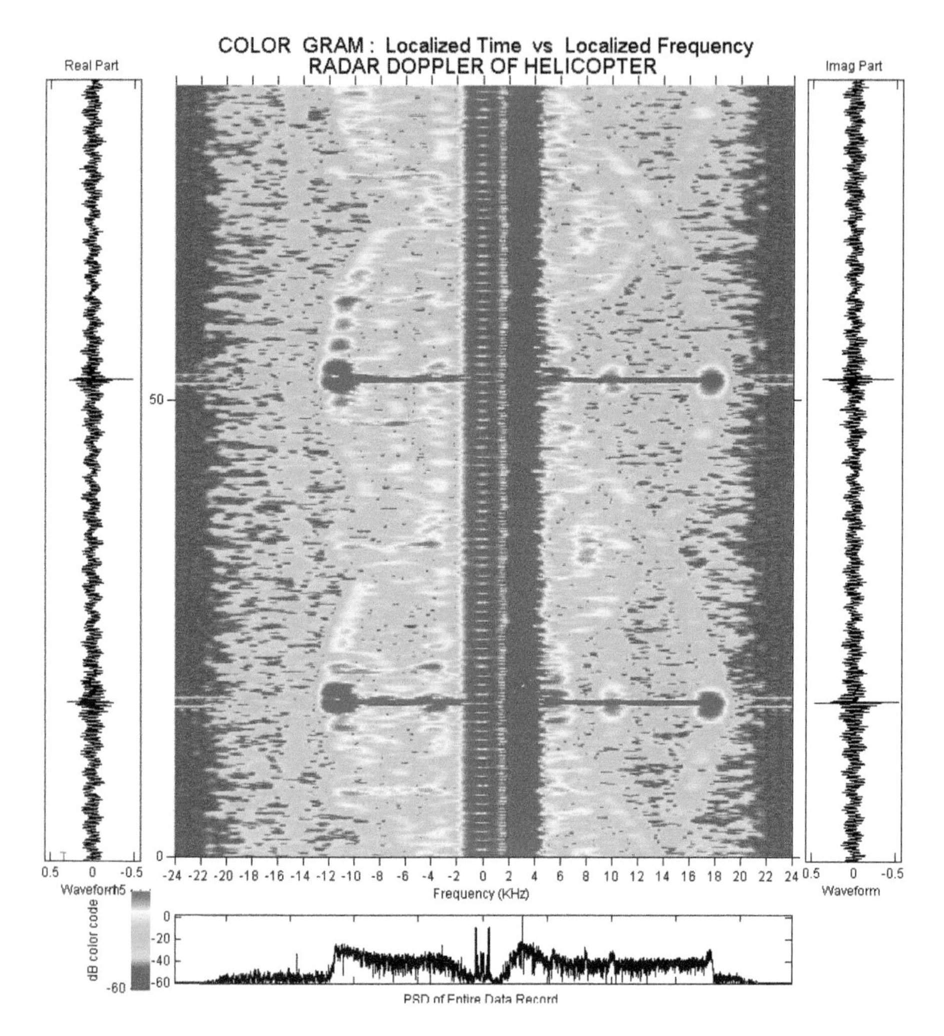

Figure 3.8: Time-frequency gram of helicopter radar doppler by periodogram analysis.

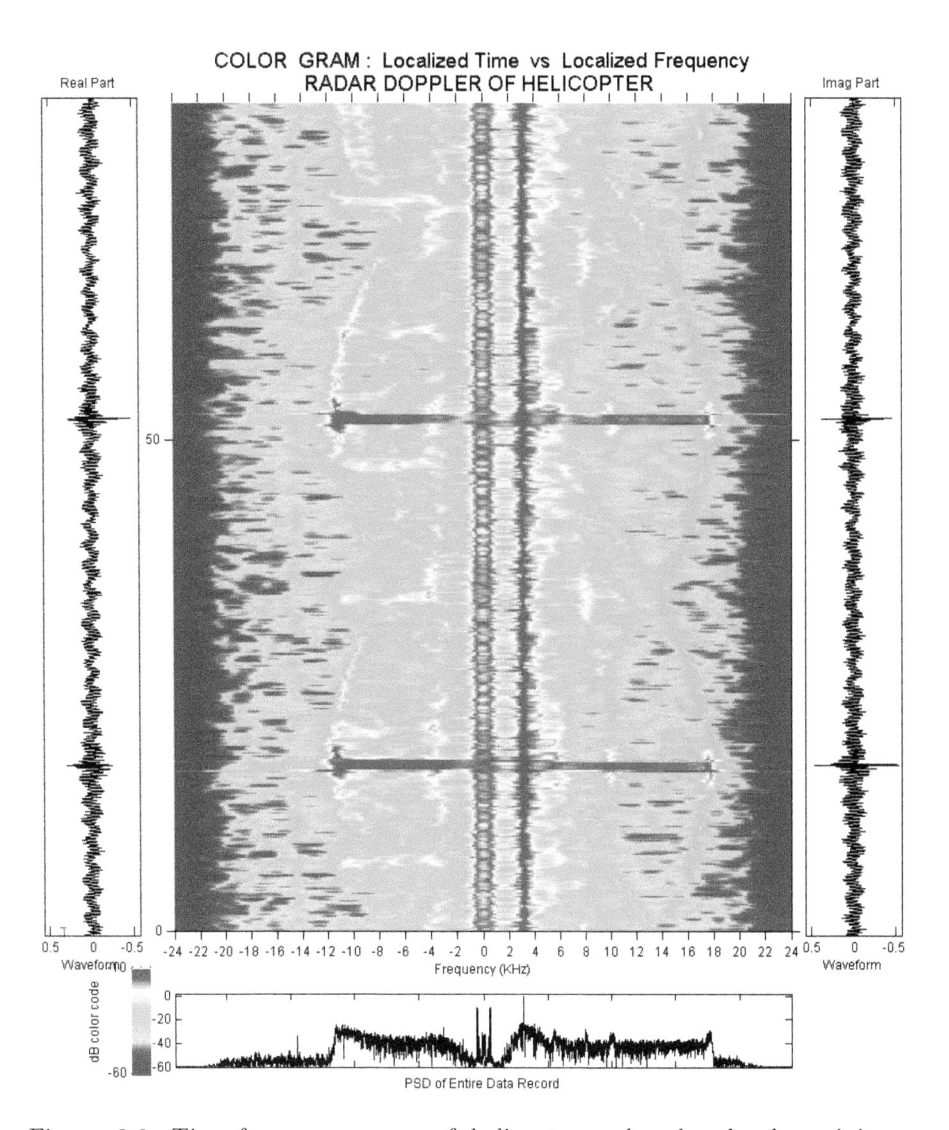

Figure 3.9: Time-frequency gram of helicopter radar doppler by minimum variance spectral estimation.

Chapter 4

TWO-CHANNEL FREQUENCY ANALYSIS

Demonstration script `spectrum_demo_MC` provides processing and plotting of three two-channel signal sources to illustrate use of multichannel spectral estimation. Multichannel spectral analysis is presented in Chapter 15 of *Digital Spectral Analysis*. Two multichannel classical (Welch, Blackman-Tukey) algorithms, three multichannel autoregressive (Yule-Walker, Vieira-Morf, Nuttall-Strand) algorithms, and the multichannel minimum variance algorithm are included in the demonstration script to provide six spectral analysis choices. Although the spectral algorithms are coded for processing M channels, the supporting plot script can only handle the two-channel case. One signal case is a real-valued two-channel AR(1) case of one hundred samples per channel that permits the reader to re-create the same case discussed in Section 15.12 of *Digital Spectral Analysis*. A second data case is a two-channel complex-valued signal of three complex sinusoidals per channel in complex white noise. Two sinusoidals are common to each channel (with time delays relative to each channel), and the third in each channel is independent from the other channel with different frequencies. The third real-valued two-channel signal case combines filtered sunspot numbers and Missouri city of St. Louis air temperatures into a two-channel signal. This data is used to prove/disprove there is modulation of earth air temperatures by the sunspot cycles.

The following script inputs will provide a demonstration of two-channel periodogram analysis and results plot of the three complex sinusoidals in white noise signal case. As seen in Fig. 4.1, the script initially produces a plot of four components: channel 1 autospectrum estimate (upper left), channel 2 autospectrum (upper right), the magnitude squared coherence between channels (lower left), and the coherence phase (lower right). After hitting the space bar, as indicated in the script below, the coherence phase plot is replaced by a plot of the relative time delay between channels, which is the final result shown in Fig. 4.1. The time delay is obtained by replacing the phase θ at each frequency f by $t_{delay} = \theta/(2\pi f)$. Having estimated the relative time delays among multiple sensors at different spatial locations can locate a common signal spatial source

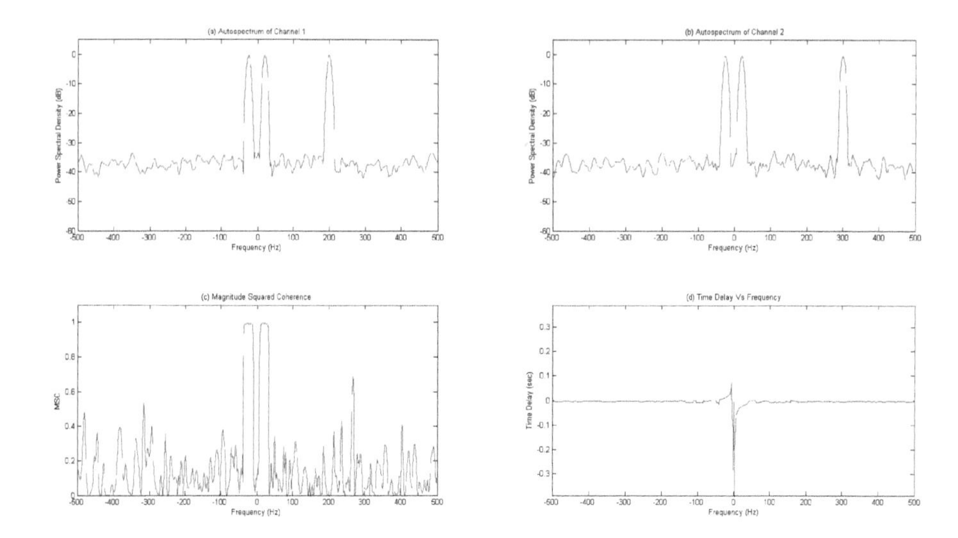

Figure 4.1: Two-channel periodogram analysis for complex sinusoidals in white noise.

received at each sensor (with differing delays to each sensor) by a process of triangular direction finding. Where the coherence magnitude is large (ideally close to 1) would indicate a coherent signal component shared by each signal channel, and the delays may be extracted at the frequencies where the coherence is large to perform a series of bearing lines, which locates the target where the lines overlap.

```
This script demonstrates MULTICHANNEL spectral analysis.

***** MULTICHANNEL DATA SOURCES *****
Data choices for 2-channel spectral analysis demonstration:
  1 -- Real-valued 2-channel AR(1) test case [Section 15.12 of course text]
  2 -- Complex-valued 2-channel 3-sinusoids [2 common, 1 not common] in
       white noise
  3 -- Filtered sunspot numbers and St Louis temperatures for years 1855 to 1968
Enter number of data source choice: 2

***** MULTICHANNEL SPECTRAL TECHNIQUE *****
Multichannel spectral analysis technique choices:
     1 -- multichannel Welch periodogram method
     2 -- multichannel Blackman-Tukey correlogram method
     3 -- multichannel autoregressive Yule-Walker method
     4 -- multichannel autoregressive Vieira-Morf (MC lattice) method
     5 -- multichannel autoregressive Nuttall-Strand (MC lattice) method
     6 -- multichannel minimum variance PSD method
Enter number of algorithm choice: 1

***** SPECTRAL PARAMETERS *****
Window choices:  0 -- none , 1 -- Hamming , 2 -- Nuttall
Enter number of window choice: 1
```

```
Enter segment size per periodogram (# samples): 128
Enter inter-segment overlap (# samples): 64

***** PLOT 2-CHANNEL RESULTS *****
Data is complex-valued.
Press any key to replace coherence phase plot with time delay estimate plot.
```

Fig. 4.1 indicates strong coherence (close to 1) at the two common sinusoidal frequencies, and no coherence at the independent frequency in each channel.

If the demonstration is rerun now for the multichannel autoregressive spectral analysis by the Nuttall-Strand algorithm using the following script entries,

```
This script demonstrates MULTICHANNEL spectral analysis.

***** MULTICHANNEL DATA SOURCES *****
Data choices for 2-channel spectral analysis demonstration:
   1 -- Real-valued 2-channel AR(1) test case [Section 15.12 of course text]
   2 -- Complex-valued 2-channel 3-sinusoids [2 common, 1 not common] in
        white noise
   3 -- Filtered sunspot numbers and St Louis temperatures for years 1855 to 1968
Enter number of data source choice: 2

***** MULTICHANNEL SPECTRAL TECHNIQUE *****
Multichannel spectral analysis technique choices:
     1 -- multichannel Welch periodogram method
     2 -- multichannel Blackman-Tukey correlogram method
     3 -- multichannel autoregressive Yule-Walker method
     4 -- multichannel autoregressive Vieira-Morf (MC lattice) method
     5 -- multichannel autoregressive Nuttall-Strand (MC lattice) method
     6 -- multichannel minimum variance PSD method
Enter number of algorithm choice: 5

***** SPECTRAL PARAMETERS *****
Enter order of multichannel autoregressive process: 12
Select: 1--forward MC AR PSD , 2--backward MC AR PSD: 1

***** PLOT 2-CHANNEL RESULTS *****
Data is complex-valued.
Press any key to replace coherence phase plot with time delay estimate plot.
```

the results are shown in Fig. 4.2. The multichannel AR spectra show strong coherence at the two common frequencies, but also some spikes at the independent frequencies where there should be no coherence. This is a leakage phenomenon characteristic of all multichannel AR algorithms. The cause is described in *Digital Spectral Analysis* Section 15.12.

An interesting two-channel real-valued pair of signals to examine for coherent linkages involves sunspot numbers. Sunspot activity on the sun's surface has long been suspected to modulate phenomena on earth, such as rain/draught cycles, magnetic declination cycles, and temperature cycles. Sunspot numbers and air temperature data from one US city are paired to determine if this coherent relationship exists. Some preparation is necessary before presenting the two-channel data for spectral analysis. Monthly sunspot numbers are provided for the period January 1700 to December 1993 for a total 3,528 samples.

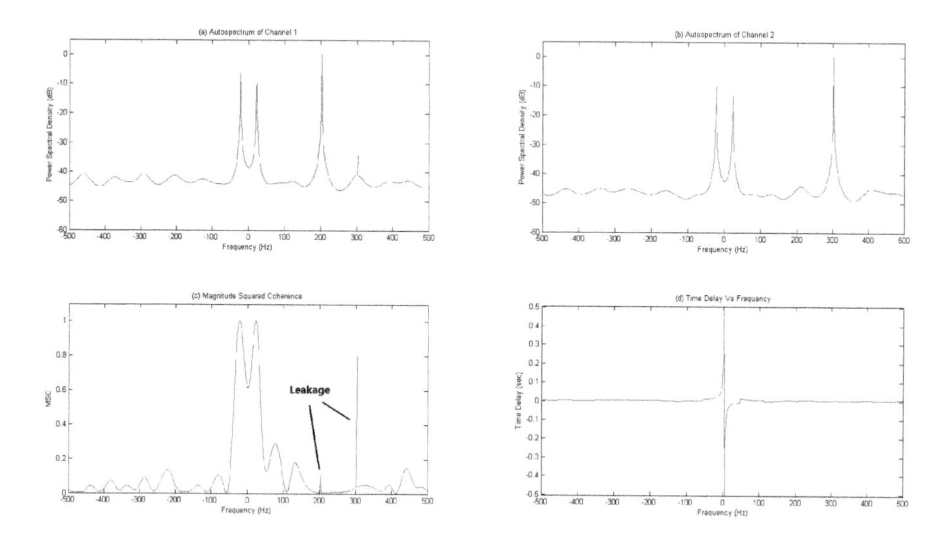

Figure 4.2: Two-channel autoregressive (Nuttall-Strand) analysis for complex sinusoidals in white noise.

Monthly average temperature for St. Louis in degrees C is provided from January 1945 to December 1978 for a total of 1,608 samples. Sunspot and temperature data are low pass filtered to remove high frequency "noise" above one cycle per year, then decimated by 6 (samples only for January and July), and finally high pass filtered to remove the strong DC component (measurements are mostly positive valued), yielding 530 half-year samples from July 1714 to January 1979 for sunspots and 228 samples from January 1855 to July 1968 (114 years) for temperature. Corresponding time-aligned 228 sunspot and temperature samples are then used to create an $x(1:228,2)$ two-channel MATLAB signal array.

The spectral results are shown in Figs. 4.3 (periodogram) and 4.4 (AR by Nuttall Strand algorithm). The only changes running the multichannel demonstration scripts from the prior scripts are the choice of source data (3), segment size 75 and segment overlap 38 in the case of periodogram analysis, and order 25 in the case of AR analysis. Neither figure shows strong coherence values (close to 1) at the strongest sunspot cycle (0.09 cycles/year or its inverse 11 years/cycle). Thus, one may conclude there is not strong evidence for a linkage of effects of sunspots on air temperature.

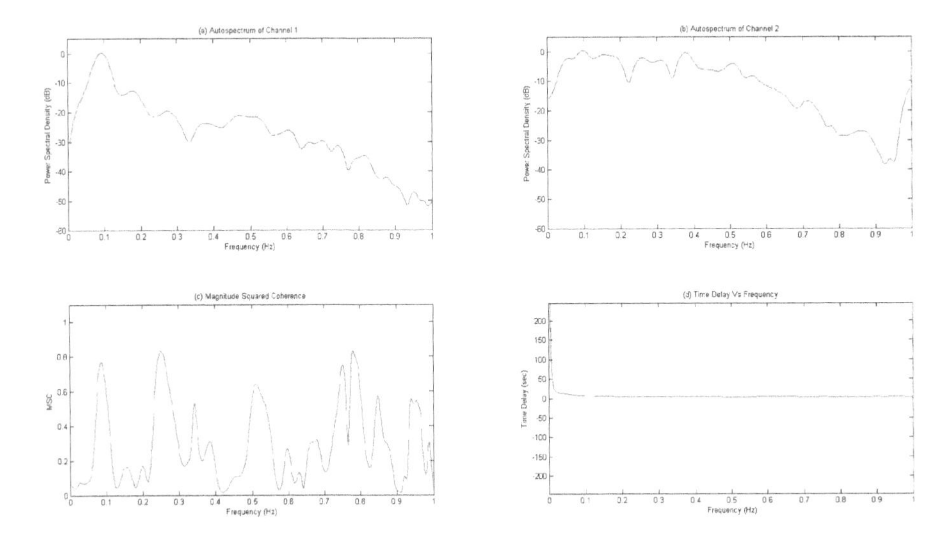

Figure 4.3: Two-channel periodogram analysis of filtered sunspot numbers and air temperature.

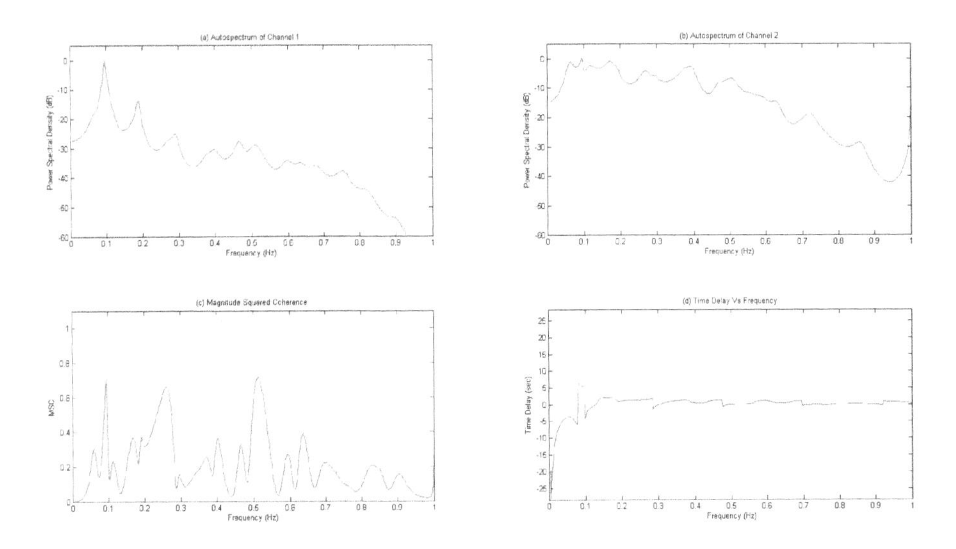

Figure 4.4: Two-channel autoregressive (Nuttall-Strand) of filtered sunspot numbers and air temperature.

Chapter 5

TWO-DIMENSIONAL DUAL FREQUENCY ANALYSIS

Two-dimensional (2-D) spectral analysis is presented in Chapter 16 of *Digital Spectral Analysis*. Demonstration script `spectrum_demo_2D` provides spectral processing and plotting of three 2-D signal sources to illustrate use of 2-D spectral estimation. Two 2-D classical (Welch, Blackman-Tukey) algorithms, two 2-D autoregressive (Yule-Walker, lattice) algorithms, and the 2-D minimum variance spectral algorithm are included in the demonstration script to provide five spectral analysis choices. There are three 2-D sources provided for the demonstration script. Two synthetic 8×8 sources contain either complex or real sinusoidals in white noise. These very small 2-D data arrays will be used to test 2-D resolution performance with limited 2-D data. They also illustrate the automatic change in frequency axis plotting depending on whether the data is real-valued or complex-valued. The third data array is an actual gray scale image with a textured pattern (see Fig. 5.1) that appears to have both low and high spatial frequency content. The image has positive-only gray scale values, so the pixel average must be calculated and removed from the data to prevent a large DC spike in the spectral estimate at (0 Hz, 0 Hz), which is performed in the preprocessing before the spectrum is estimated.

The use of the 2-D demonstration script will first be illustrated by a 2-D periodogram estimate of the 8×8 complex sinusoids in white noise case. This is a limited data case in which the resolution capability is being tested. In 1-D, the periodogram was estimated by averaging sample periodograms of segments. In 2-D, the final periodogram is estimated by averaging sample 2-D periodograms of subarrays, with definitions of the subarray given in Fig. 5.2. The entries to the 2-D demonstration script for this case are:

Figure 5.1: Texture image used for 2-D spectral analysis to determine spatial frequency content.

```
This script demonstrates 2-D spectral analysis techniques.

***** TWO-DIMENSIONAL DATA SOURCE *****
2-D data choices for two-dimensional spectral analysis demonstration:
    1 -- 8 x 8 2-D signal: three 2-D complex sines + 2-D white noise
    2 -- 423 x 427 texture image
    3 -- 8 x 8 2-D signal: two 2-D real sines + additive noise
Enter number of 2-D data source choice: 1

***** 2-D SPECTRAL ANALYSIS TECHNIQUE *****
Two-dimensional spectral analysis technique choices:
     1 -- two-dimensional Welch periodogram method
     2 -- two-dimensional Blackman-Tukey correlogram method
     3 -- two-dimensional autoregressive Yule-Walker method
     4 -- two-dimensional autoregressive lattice method
     5 -- two-dimensional minimum variance PSD method
Enter number of algorithm choice: 1

***** 2-D SPECTRAL ANALYSIS PARAMETERS *****
Enter row subarray dimension per periodogram (#samples): 8
Enter column subarray dimension per periodogram (#samples): 8
Enter inter-subarray overlap along rows (#samples): 0
Enter inter-subarray overlap along columns (#samples): 0
Enter window choice-- 0 [None], 1 [Hamming], 2 [Nuttall]: 1
```

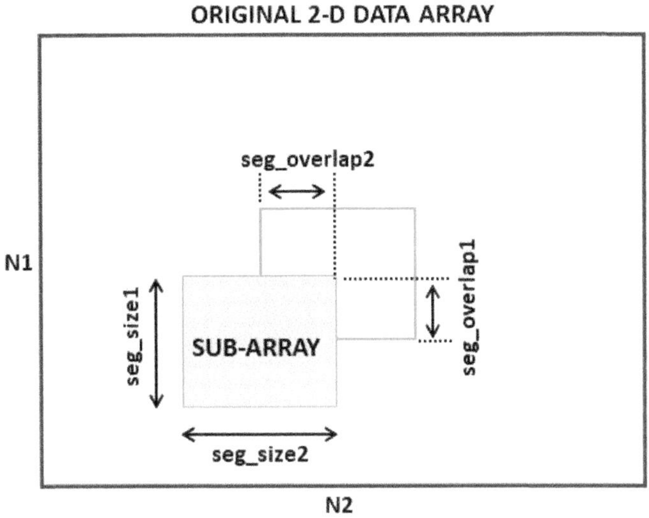

Figure 5.2: Parameter definitions to form 2-D subarrays for averaging 2-D sub-array periodograms (extension of concept of 1-D segments for averaging 1-D periodograms).

```
***** COMPUTING 2-D SPECTRUM BASED ON SELECTIONS *****

***** PLOT 2-D SPECTRAL RESULTS *****
Data is complex-valued.
2-D spectral analysis will be scaled in logarithmic units (dB).
Logarithmic range in dB that 2-D PSD is to be plotted (typical range is
  40 to 80 dB): 50
Enter text for 2-D spectral analysis title (no quotes around text):
  3 COMPLEX SINES + WHITE NOISE
```

The plotted results in both 3-D and 2-D renderings are shown in Fig. 5.3. A color bar adjacent to the right of the 2-D plot shows the magnitude range in dB that is plotted in color. Note that the frequency axis adjusts scale to the given sampling interval inputs provided in the demonstration script, and also adjusts for real-valued vs complex-valued data. Note also that the 2-D periodogram is unable to resolve all three sinusoidals in the very limited 2-D signal.

The same data can be processed by an alternative 2-D spectral estimator to achieve resolution of all three sinusoidals. If the demonstration script is rerun for the 2-D AR (lattice algorithm) spectral estimate

```
This script demonstrates 2-D spectral analysis techniques.

***** TWO-DIMENSIONAL DATA SOURCE *****
2-D data choices for two-dimensional spectral analysis demonstration:
    1 -- 8 x 8 2-D signal: three 2-D complex sines + 2-D white noise
    2 -- 423 x 427 texture image
```

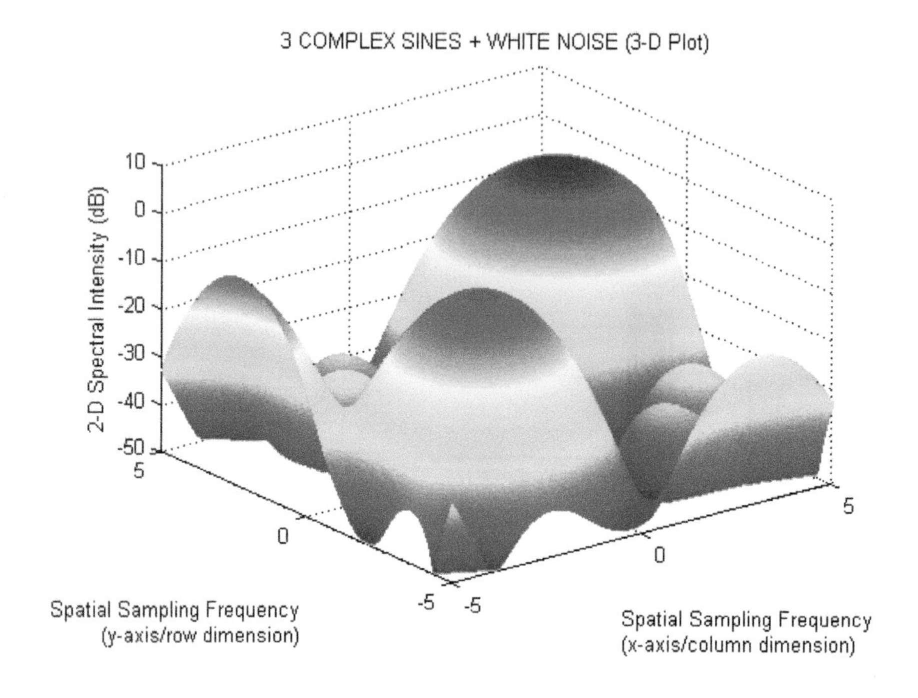

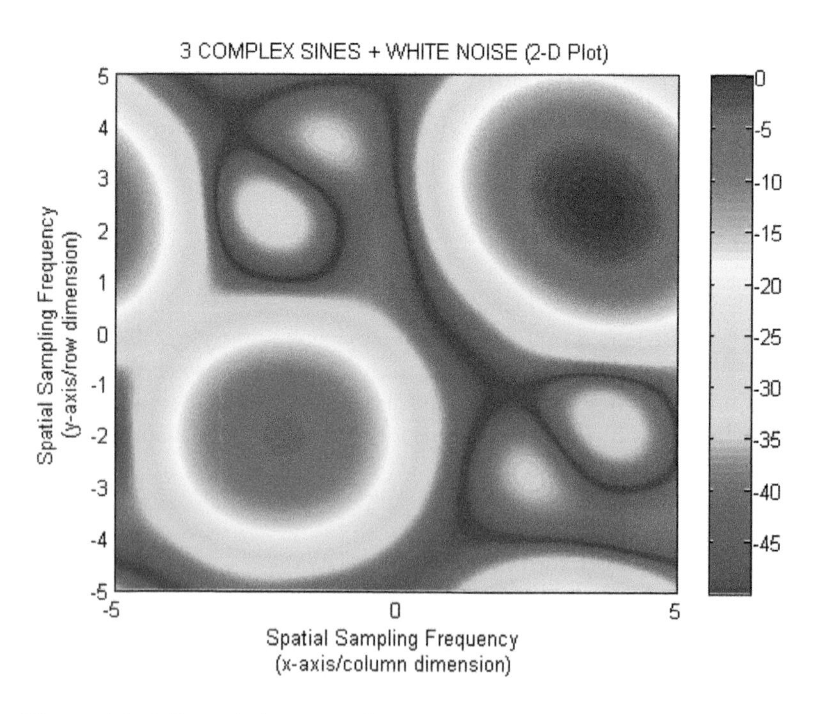

Figure 5.3: 2-D periodogram of 8×8 complex sinusoids in white noise.

```
    3 -- 8 x 8 2-D signal: two 2-D real sines + additive noise
Enter number of 2-D data source choice: 1

***** 2-D SPECTRAL ANALYSIS TECHNIQUE *****
Two-dimensional spectral analysis technique choices:
     1 -- two-dimensional Welch periodogram method
     2 -- two-dimensional Blackman-Tukey correlogram method
     3 -- two-dimensional autoregressive Yule-Walker method
     4 -- two-dimensional autoregressive lattice method
     5 -- two-dimensional minimum variance PSD method
Enter number of algorithm choice: 4

***** 2-D SPECTRAL ANALYSIS PARAMETERS *****
Enter row order of two-dimensional autoregressive process: 3
Enter column order of two-dimensional autoregressive process: 3

***** COMPUTING 2-D SPECTRUM BASED ON SELECTIONS *****

***** PLOT 2-D SPECTRAL RESULTS *****
Data is complex-valued.
2-D spectral analysis will be scaled in logarithmic units (dB).
Logarithmic range in dB that 2-D PSD is to be plotted (typical range is
  40 to 80 dB): 50
Enter text for 2-D spectral analysis title (no quotes around text):
  AR SPECTRUM FOR 3 SINES + NOISE
```

the results are shown in Fig. 5.4. Note that, as in the 1-D case, the AR algorithm is able to resolve signal components that the classical methods, such as the periodogram, cannot.

The gray scale texture image frequency content can be obtained by the following demonstration script entries using the 2-D minimum variance spectral estimator:

```
This script demonstrates 2-D spectral analysis techniques.

***** TWO-DIMENSIONAL DATA SOURCE *****
2-D data choices for two-dimensional spectral analysis demonstration:
    1 -- 8 x 8 2-D signal: three 2-D complex sines + 2-D white noise
    2 -- 423 x 427 texture image
    3 -- 8 x 8 2-D signal: two 2-D real sines + additive noise
Enter number of 2-D data source choice: 2

***** 2-D SPECTRAL ANALYSIS TECHNIQUE *****
Two-dimensional spectral analysis technique choices:
     1 -- two-dimensional Welch periodogram method
     2 -- two-dimensional Blackman-Tukey correlogram method
     3 -- two-dimensional autoregressive Yule-Walker method
     4 -- two-dimensional autoregressive lattice method
     5 -- two-dimensional minimum variance PSD method
Enter number of algorithm choice: 5

***** 2-D SPECTRAL ANALYSIS PARAMETERS *****
Enter row order of 2-D adaptive minvar filter: 10
Enter column order of 2-D adaptive minvar filter: 10
```

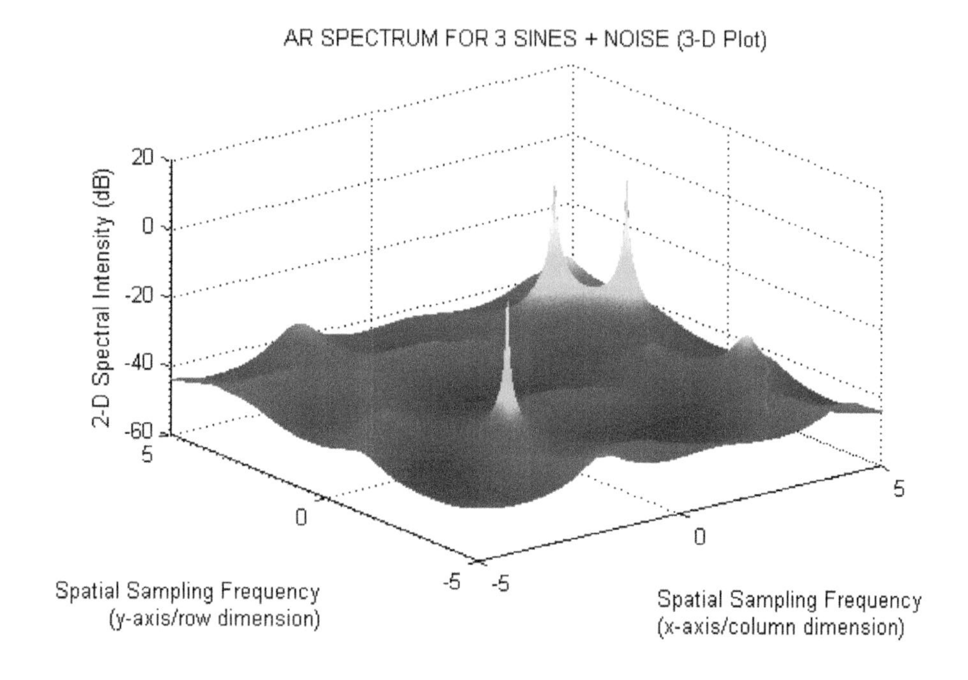

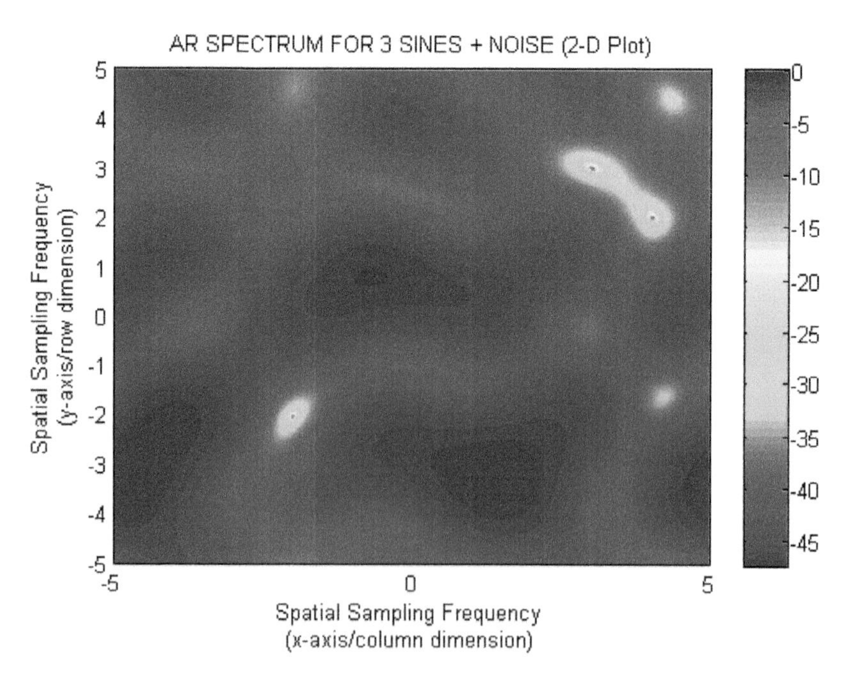

Figure 5.4: 2-D autoregressive [lattice] spectrum of 8×8 complex sinusoids in white noise.

```
***** COMPUTING 2-D SPECTRUM BASED ON SELECTIONS *****

***** PLOT 2-D SPECTRAL RESULTS *****
Data is real-valued.
2-D spectral analysis will be scaled in logarithmic units (dB).
Logarithmic range in dB that 2-D PSD is to be plotted (typical range is
  40 to 80 dB): 50
Enter text for 2-D spectral analysis title (no quotes around text):
  MINIMUM VARIANCE SPECTRUM OF TEX
```

The results are shown in Fig. 5.5. Note that there are both low spatial frequency and high spatial frequency features, but negligible middle frequency content.

The real power of the alternative 2-D spectral methods may be seen in more complicated 2-D signals. The application to synthetic aperture radar (SAR) using complex-valued "phase history" is such a case. This data is of very low signal to noise ratio (SNR) and a spectral analysis is applied to form a computed radar image from the phase history. Creating a cartesian set of 2-D sample values from the radar collection system, typically acquired from an aircraft or drone, is a detailed processing story beyond the scope of this user guide. Current digital SAR systems normally use a 2-D FFT with windowing to compute the final processing step that yields a computed complex image from the 2-D phase history. Taking the squared magnitude of the complex-valued FFT then yields the viewable radar spatial image. These two final steps are equivalent to calculating a 2-D sample periodogram. Using as an alternative 2-D minimum variance spectral analysis can greatly enhance the radar image. Consider in Fig. 5.6 the radar image created by applying an FFT to a 256 x 256 grid of phase history samples. The resulting radar image shows a tall structure in the scene that is likely an advertisement sign with two decorative structures on either side, as might be used at the entry of an industrial development area. The direction of the radar illumination source is inferred from the shadowing behind the tall structures.

Using a 2-D minimum variance spectral estimator of size 50 x 50 parameters is applied to the complex-valued phase history, rather than a 2-D FFT, the result is shown in Fig. 5.7. Note the sharpened features in the radar image and the now clearly distinguishable small fence posts that can be seen that were not quite resolvable by the 2-D FFT processing.

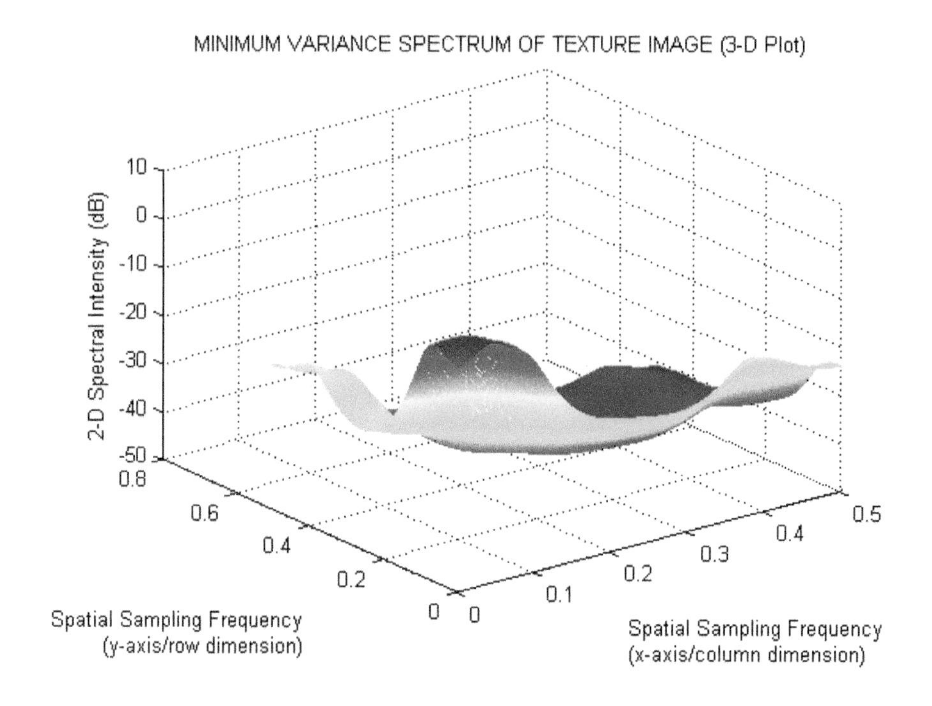

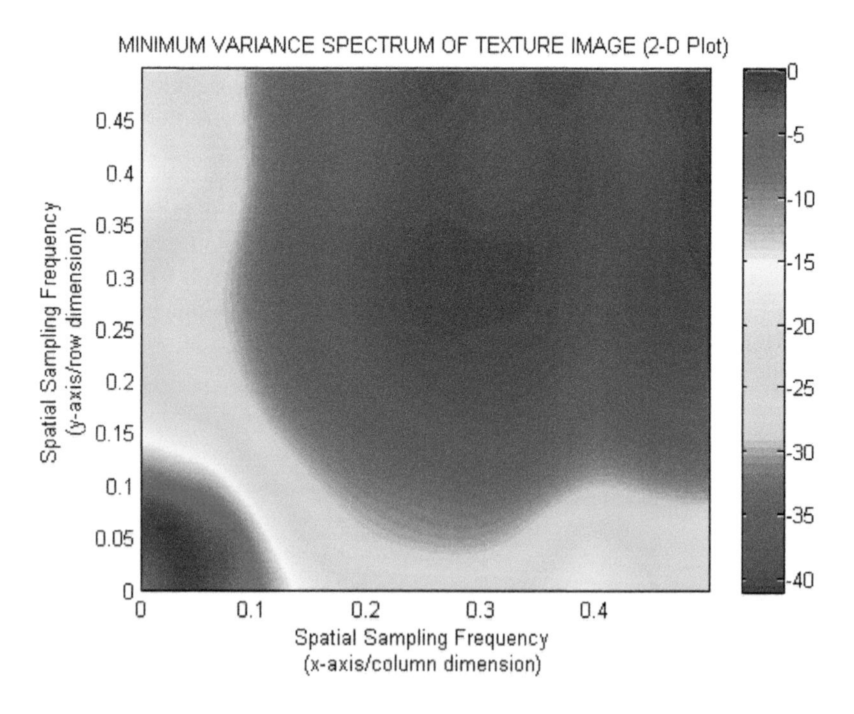

Figure 5.5: 2-D minimum variance spectrum of texture image pixels.

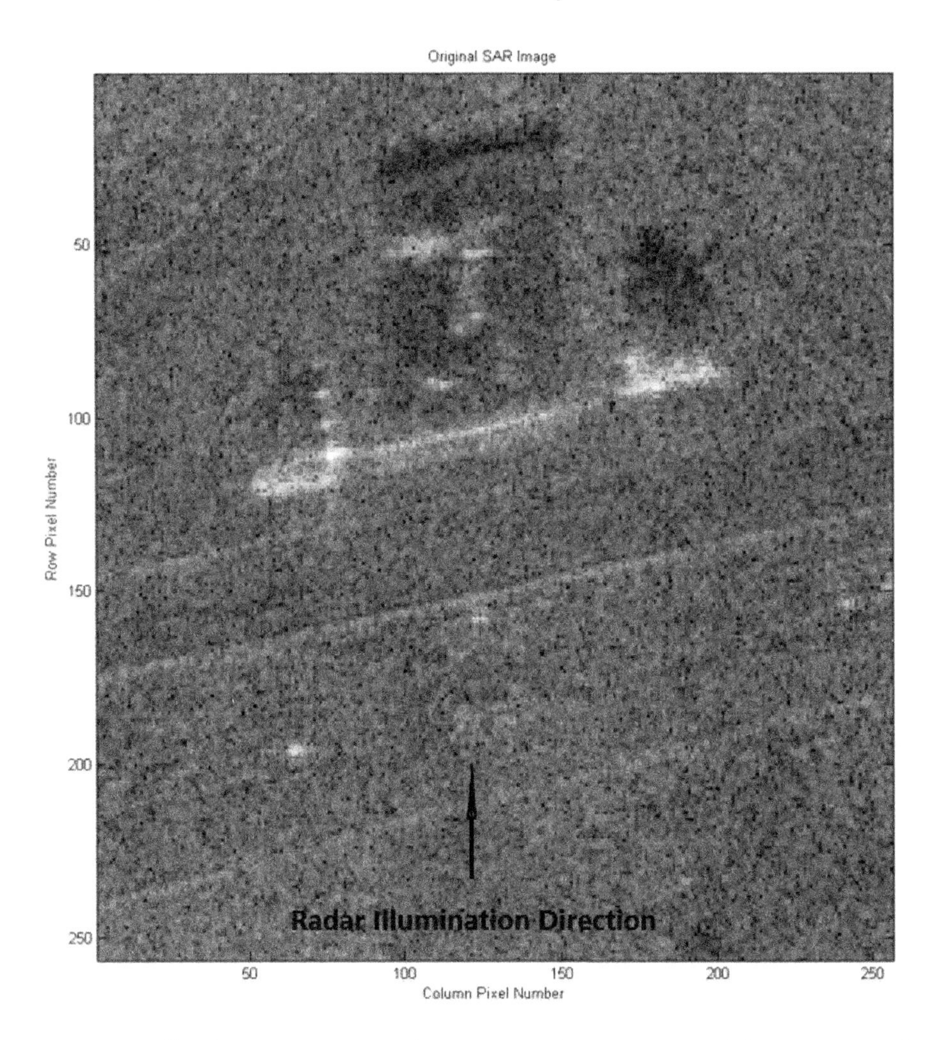

Figure 5.6: Synthetic aperture radar computed image created from magnitude of 2-D FFT applied to complex radar phase history.

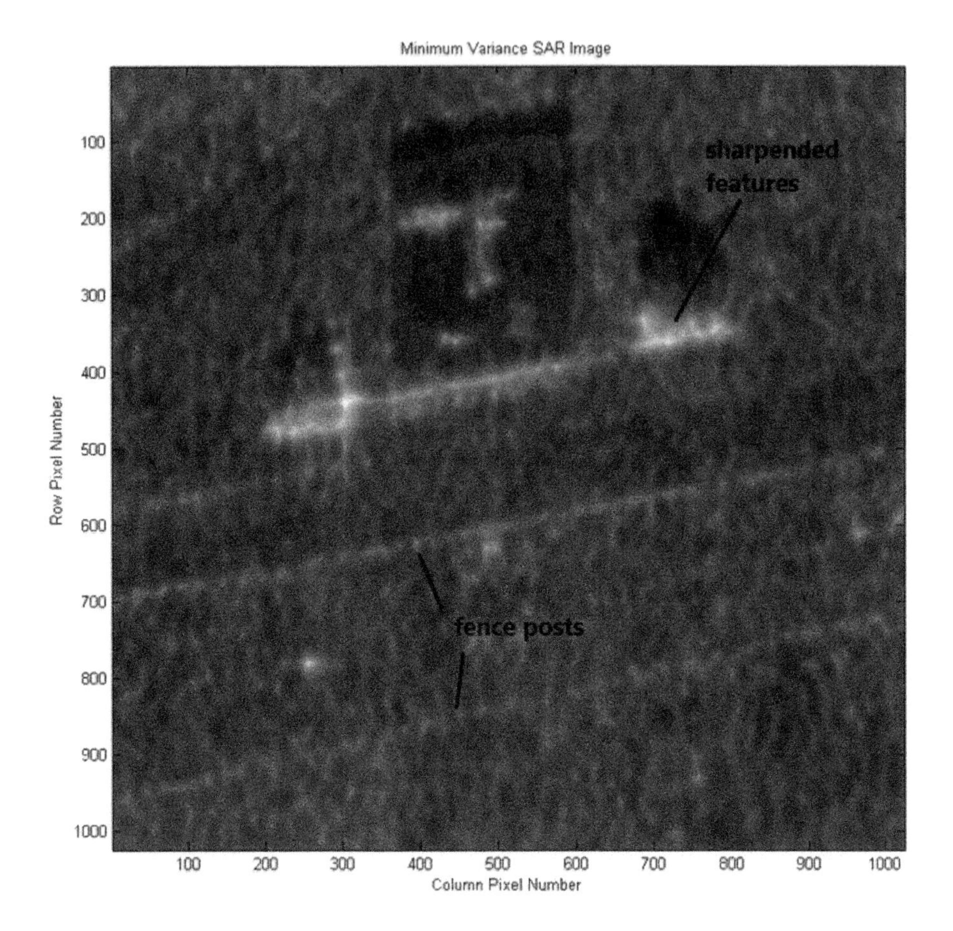

Figure 5.7: Synthetic aperture radar computed image created by 2-D minimum variance spectral algorithm.

Chapter 6

MATLAB DIGITAL SPECTRAL ANALYSIS FUNCTIONS

This chapter is an alphabetical organization of every MATLAB function developed for demonstration scripts covered in Chapters 1 to 5 of this user guide and referenced in the text *Digital Spectral Analysis*. Each function is briefly described and a source listing is also provided.

ar_psd_MC

Given M signal channels, this function accepts the order p estimated forward and backward $M \times M$ AR parameter matrices and the $M \times M$ estimated forward and backward white noise variance matrices provided by MATLAB functions `covar_lp_MC`, `lattice_MC`, or `yule_walker_MC` to compute one of two possible $M \times M$ auto- and cross-power spectral density (PSD) matrices per frequency, depending on the setting of `f_or_b` (see *Digital Spectral Analysis* Section 15.10.5)

$$\mathbf{PSD}^f_{\mathrm{AR}}(f_k) = \begin{pmatrix} P^f_{11}(f_k) & \dots & P^f_{1M}(f_k) \\ \vdots & \ddots & \vdots \\ P^f_{M1}(f_k) & \dots & P^f_{MM}(f_k) \end{pmatrix} = T_s\big[\mathbf{A}(f_k)\big]^{-1}\mathbf{P}^a\big[\mathbf{A}(f_k)\big]^{-H},$$

or

$$\mathbf{PSD}^b_{\mathrm{AR}}(f_k) = \begin{pmatrix} P^b_{11}(f_k) & \dots & P^b_{1M}(f_k) \\ \vdots & \ddots & \vdots \\ P^b_{M1}(f_k) & \dots & P^b_{MM}(f_k) \end{pmatrix} = T_s\big[\mathbf{B}(f_k)\big]^{-1}\mathbf{P}^b\big[\mathbf{B}(f_k)\big]^{-H}$$

in which T_s is the sampling interval (sec) and

$$\mathbf{A}(f_k) = \mathbf{I} + \sum_{n=1}^{p} \mathbf{A}[n]\exp(-j2\pi f_k n T_s)$$

$$\mathbf{B}(f_k) = \exp(-j2\pi f_k p T_s)\left[\mathbf{I} + \sum_{n=1}^{p} \mathbf{B}[n]\exp(j2\pi f_k n T_s)\right].$$

Function `ar_psd_MC` accepts the autoregressive (AR) multichannel parameter matrices as block row vectors of p matrix elements

$$\mathbf{a}_p = \begin{pmatrix}\mathbf{A}[1] & \dots & \mathbf{A}[p]\end{pmatrix}$$
$$\mathbf{b}_p = \begin{pmatrix}\mathbf{B}[p] & \dots & \mathbf{B}[1]\end{pmatrix},$$

with scalar dimension M by $M * p$. The scalar dimension of the computed `psd` block column vector is $M *$ `num_psd` by M. The frequency f_k in Hz associated with the PSD matrix k indexing is $f_k = (k - \texttt{num\_psd}/2)/(\texttt{num\_psd} * T_s)$ for $k = 1, \dots,$ `num_psd`.

```
function psd=ar_psd_MC(f_or_b,num_psd,Ts,Pa,a,Pb,b)
% Copyright (c) 2019 by S. Lawrence Marple Jr.

% This function computes the multichannel AR power spectral density matrix
% over the specified number of frequency points using the forward or backward
% multichannel linear prediction (LP)/AR parameter block row vectors
%
%        psd = ar_psd_MC(f_or_b,num_psd,Ts,Pa,a,Pb,b)
%
```

```
% f_or_b  -- select: 1--forward PSD computation, 2--backward PSD computation
% num_psd -- number of frequency samples in spectrum
% Ts      -- sample interval in appropriate units (seconds, meters, etc.)
% Pa      -- forward linear prediction/autoregressive white noise variance
% a       -- block vector of forward linear prediction / AR parameters
% Pb      -- backward linear prediction/autoregressive white noise variance
% a       -- block vector of backward linear prediction / AR parameters
% psd     -- block column vector of multichannel PSD matrices

[num_chs,n] = size(a);
p = n/num_chs;        % order of multichannel AR process
psd = complex(zeros(num_chs*num_psd,num_chs));    %preallocate memory
index = 1:num_chs;

if f_or_b == 1        % forward multichannel AR PSD computation

    a = [eye(num_chs) a];
    A = complex(zeros(num_chs,num_chs*num_psd));     % preallocate memory
    for i=1:num_chs
        for j=1:num_chs
            A(i,j:num_chs:j+(num_psd-1)*num_chs) = ...
              fftshift(fft(a(i,j:num_chs:j+p*num_chs),num_psd));
        end
    end
    for i=1:num_psd
        c = A(:,index);
        psd(index,:) = Ts*(c\Pa/c');
        index = index + num_chs;
    end

else                  % backward multichannel AR PSD computation (default)

    b = [b eye(num_chs)];
    B = complex(zeros(num_chs,num_chs*num_psd));      % preallocate memory
    for i=1:num_chs
        for j=1:num_chs
            B(i,j:num_chs:j+(num_psd-1)*num_chs) = ...
              fftshift(conj(fft(conj(b(i,j+p*num_chs:-num_chs:j)),num_psd)));
        end
    end
    for i=1:num_psd
        c = B(:,index);
        psd(index,:) = Ts*(c\Pb/c');
        index = index + num_chs;
    end

end
```

ar_psd_2D

Given a 2-D signal for which a 2-D autoregressive (AR) model of order (p_1, p_2) has been generated, this function accepts the estimated first and fourth quarter plane AR parameter matrices of dimension $(p_1+1) \times (p_2+1)$ and their associated scalar white noise variances. These estimates, which are provided by MATLAB functions `lattice_2D` or `yule_walker_2D`, are used to compute one of three possible matrices of 2-D power spectral density (PSD), depending on the setting of `psd_type` (see *Digital Spectral Analysis* Section 16.7.5).

First (Q1), second (Q2), third (Q3) and fourth (Q4) quadrant AR parameter arrays $a_1[m,n]$, $a_2[m,n]$, $a_3[m,n]$, and $a_4[m,n]$ are defined over the indicated indexing ranges

$$a[m,n] = \begin{cases} a_1[m,n] & 0 \leq m \leq p_1 (1 \leq n \leq p_2), 1 \leq m \leq p_1 (n=0) \\ a_2[m,n] & -p_1 \leq m \leq 0 (1 \leq n \leq p_2), -p_1 \leq m \leq -1 (n=0) \\ a_3[m,n] & -p_1 \leq m \leq 0 (-p_2 \leq n \leq -1), -p_1 \leq m \leq -1 (n=0) \\ a_4[m,n] & 0 \leq m \leq p_1 (-p_2 \leq n \leq -1), 1 \leq m \leq p_1 (n=0) \end{cases} .$$

Since the Q2 AR parameters are the complex conjugates of the Q4 AR parameters, and the Q3 AR parameters are the complex conjugates of the Q1 AR parameters, only two quarter plane AR parameters are unique for purposes of computing the 2-D AR PSD values. This function then generates the `psd` 2-D values as

$$\begin{aligned} &\mathbf{PSD}_{\mathrm{Q1}}(f_1(k), f_2(l)) \\ &= T_1 T_2 \frac{p_{Q1}}{\left| \sum_{m=0}^{p_1} \sum_{n=0}^{p_2} a_1[m,n] \exp(-j2\pi[f_1(k) m T_1 + f_2(l) n T_2]) \right|^2} \end{aligned}$$

and

$$\begin{aligned} &\mathbf{PSD}_{\mathrm{Q4}}(f_1(k), f_2(l)) \\ &= T_1 T_2 \frac{p_{Q4}}{\left| \sum_{m=0}^{p_1} \sum_{n=-p_2}^{0} a_4[m,n] \exp(-j2\pi[f_1(k) m T_1 + f_2(l) n T_2]) \right|^2} \end{aligned}$$

in which scalars p_{Q1} and p_{Q4} are the Q1 and Q4 quarter plane white noise variances. An unbiased PSD response is obtained by combining the Q1 and Q4 PSDs

$$\frac{1}{\mathbf{PSD}_{\mathrm{comb}}(f_1(k), f_2(l))} = \frac{1}{\mathbf{PSD}_{\mathrm{Q1}}(f_1(k), f_2(l))} + \frac{1}{\mathbf{PSD}_{\mathrm{Q4}}(f_1(k), f_2(l))}$$

and is recommended for most applications. Each quadrant AR parameter matrix is of dimension (p_1+1) by (p_2+1). The function output `psd` has dimension **num_psd1** by **num_psd2**. The row frequency indexing of the 2-D **psd** matrix is $f_1(k) = (k - \texttt{num\_psd1}/2)/(\texttt{num\_psd1} * T_1)$ for $k = 1, \ldots, \texttt{num\_psd1}$. The column frequency indexing $f_2(l) = (l - \texttt{num\_psd2}/2)/(\texttt{num\_psd2} * T_2)$ for $l = 1, \ldots, \texttt{num\_psd2}$.

```
function psd=ar_psd_2D(psd_type,num_psd1,num_psd2,T1,T2,p_Q1,a_Q1,p_Q4,a_Q4)
% Copyright (c) 2019 by S. Lawrence Marple Jr.

% Generate the first, fourth, or combined quarter-plane two-dimensional AR
% power spectral density.
%
%     psd = ar_psd_2D(psd_type,num_psd1,num_psd2,T1,T2,p_Q1,a_Q1,p_Q4,a_Q4)
%
% psd_type -- select: 1--Q1 plane PSD, 2--Q4 plane PSD, 3--combined Q1 & Q4
% num_psd1 -- number of row PSD values to evaluate (power of two)
% num_psd2 -- number of column PSD values to evaluate (power of two)
% T1       -- sample interval along rows in seconds
% T2       -- sample interval along columns in seconds
% p_Q1     -- first-quadrant AR white noise variance
% a_Q1     -- two-dimensional Q1-plane AR parameter array
% p_Q4     -- fourth-quadrant AR white noise variance
% a_Q4     -- two-dimensional Q4-plane AR parameter array
% psd      -- two-dimensional quarter-plane AR power spectral density matrix

% two-dimensional Q1-plane AR power spectral density
if  psd_type ~= 2
    xfm = fft2(a_Q1,num_psd1,num_psd2);
    psd = real(xfm).^2 + imag(xfm).^2;
    if psd_type == 1
        psd = fftshift(T1*T2*p_Q1*ones(num_psd1,num_psd2)./psd); % Q1 plane
        return
    else
        psd_Q1 = psd/(T1*T2*p_Q1);                                % combined
    end
end

% two-dimensional Q4-plane AR power spectral density
if  psd_type ~= 1
    [p1,p2] = size(a_Q4);
    a4 = zeros(num_psd1,p2);
    a4 = a_Q4(1,:);
    a4(num_psd1:-1:num_psd1-p1+2,:) = a_Q4(2:p1,:);
    xfm = fft2(a4,num_psd1,num_psd2);
    psd = real(xfm).^2 + imag(xfm).^2;
    if  psd_type == 2
        psd = fftshift(T1*T2*p_Q4*ones(num_psd1,num_psd2)./psd);    % Q4 plane
        return
    else
        psd_Q4 = psd/(T1*T2*p_Q4);
        psd = fftshift(ones(num_psd1,num_psd2)./(psd_Q1 + psd_Q4));% combined
    end
end
```

arma

This function sub-optimally estimates an ARMA(p,q) model fit by, first, estimating the autoregressive (AR) parameters and then, second, estimating the moving average (MA) parameters. The estimation procedure is presented in *Digital Spectral Analysis* Section 10.4, and uses a number of fast computational algorithms. Since an ARMA power spectral density is most sensitive to the AR parameters and order, it is estimated first. A summary of the estimation procedure:

- Estimate autocorrelation (ACS) lags out to m, in which $m > p + q$
- Perform a least squares fit of the high-order ACS lags ($> p + q$) using high-order Yule-Walker equations; this yields the AR parameters
- Filter the original data with an inverse filter formed from the AR parameters; filter output is approximately an MA process
- Use `ma` function to estimate the MA parameters and white noise variance from the filtered data

The AR and MA parameters are stored as column vectors a (dimension p) and b (dimension q) respectively.

```
function [rho,a,b]=arma(p,q,x,m)
% Copyright (c) 2019 by S. Lawrence Marple Jr.

% Estimates the AR and MA parameters of an ARMA process from sample data.
%
%      [rho,a,b] = arma(p,q,x,m)
%
% p    -- order of AR portion of ARMA model
% q    -- order of MA portion of ARMA model
% x    -- vector of data samples
% m    -- maximum ACS lag for modified Yule-Walker solution for AR (set m > p+q)
% rho  -- estimated driving white noise variance
% a    -- vector of AR parameter estimates
% b    -- vector of MA parameter estimates

if m < p+q+1
    error(' Higher lag required so solution not singular.')
end
r = correlation_sequence(m,'unbiased',x);
[rho_a,a,rho_b,a_b] = covariance_lp(p,r(m+q-p+2:2*m+1));
y = toeplitz(x(p+1:length(x)),x(p+1:-1:1))*[1;a];
[rho,b] = ma(q,y);
```

arma_psd

This function generates the power spectral density psd values for an autoregressive moving average signal, moving average signal or autoregressive signal (see *Digital Spectral Analysis* Section 6.3)

$$\mathtt{psd}_{\mathrm{ARMA}}(f_k) = T\rho \left| \frac{B(f_k)}{A(f_k)} \right|^2, \mathtt{psd}_{\mathrm{MA}}(f_k) = T\rho \left| B(f_k) \right|^2, \mathtt{psd}_{\mathrm{AR}}(f_k) = \frac{T\rho}{\left| A(f_k) \right|^2}$$

in which the Fourier transforms of the estimated AR and MA parameters are defined as

$$A(f_k) = 1 + \sum_{n=1}^{p} a[n] \exp(-j2\pi f_k nT)$$

$$B(f_k) = 1 + \sum_{n=1}^{q} b[n] \exp(-j2\pi f_k nT).$$

and T is the scalar sampling interval (sec) and ρ is the scalar estimated white noise variance. For just an MA signal, set AR parameter vector to the null vector $a = [\,]$. For just an AR signal, set MA parameter vector to the null vector $b = [\,]$. AR parameter column vector a has dimension p. MA parameter column vector b has dimension q. The psd row vector has dimension num_psd in which the frequency (Hz) is $f_k = (k - \mathtt{num\_psd}/2)/(\mathtt{num\_psd} * T)$ for $k = 1, \ldots, \mathtt{num\_psd}$.

```
function psd=arma_psd(num_psd,T,rho,a,b)
% Copyright (c) 2019 by S. Lawrence Marple Jr.

% Generates the AR, MA, or ARMA power spectral density values.
% ARMA psd: eq (6.8); MA psd: eq (6.12); AR psd: eq (6.14)
%
%     psd = arma_psd(num_psd,T,rho,a,b)
%
% num_psd -- size of psd vector (must be power of two)
% T       -- sample interval in seconds
% rho     -- variance of driving white noise process
% a       -- vector of autoregressive parameters
% b       -- vector of moving average parameters
% psd     -- vector of power spectral density values

h = fft([1;b;0],num_psd)./fft([1;a;0],num_psd);
psd = fftshift(rho*T*(real(h).^2 + imag(h).^2));
```

bilineqn

This is a supporting function for the multichannel AR parameter matrix estimation. The algorithm, described in *Digital Spectral Analysis* Section 15.10.3, will solve the bilinear matrix equation

$$\mathbf{AX} + \mathbf{XB} = \mathbf{C},$$

for $\mathbf{X}$, in which $\mathbf{A}$, $\mathbf{B}$, $\mathbf{C}$, and $\mathbf{X}$ are all $M \times M$ square real-valued or complex-valued matrices.

```
function X=bilineqn(A,B,C)
% Copyright (c) 2019 by S. Lawrence Marple Jr.

% Solves the bilinear matrix equation  AX + XB = C  based on solution given as
% text equations (15.100),(15.101).
%
%          X = bilineqn(A,B,C)
%
% A -- square matrix
% B -- square matrix
% C -- square matrix
% X -- square solution matrix

%********************** Initialization ************************
[m,n] = size(A);
d = 1;
D = eye(m,m);
P = C*B^(m-1);
Q = B^m;
factor = 1;

%********************** Main Recursion ***********************
for k = 1:m
    D = A*D;
    d = -trace(D)/k;
    factor = -1*factor;
    Q = Q + factor*d*B^(m-k);
    if k < m
        D = D + d*eye(size(D));
        P = P + factor*D*C*B^(m-1-k);
    end
end
X=P/Q;
```

correlation_sequence

This function estimates the biased autocorrelation from samples of one signal in column vector x

$$r_{xx}[m] = \begin{cases} \frac{1}{N} \sum_{n=0}^{N-m-1} x[n+m]x^*[n] & \text{for } 0 \leq m \leq N-1 \\ \frac{1}{N} \sum_{n=0}^{N-|m|-1} x^*[n+|m|]x[n] & \text{for } -(N-1) \leq m < 0 \end{cases}$$

or the biased cross correlation from samples of two signals in column vectors x and y

$$r_{xy}[m] = \begin{cases} \frac{1}{N} \sum_{n=0}^{N-m-1} x[n+m]y^*[n] & \text{for } 0 \leq m \leq N-1 \\ \frac{1}{N} \sum_{n=0}^{N-|m|-1} x[n]y^*[n+|m|] & \text{for } -(N-1) \leq m < 0 \end{cases} .$$

The unbiased autocorrelation or cross correlation estimate uses the front factor $1/(N-|m|)$, rather than the factor $1/N$ (see *Digital Spectral Analysis* Section 5.5). The estimated correlation r is stored in a column vector of dimension $2 * \texttt{max\_lag} + 1$. Note the association of lag index to the MATLAB vector index in the function commentary.

```
function r=correlation_sequence(max_lag,bias,x,y)
% Copyright (c) 2019  by S. Lawrence Marple Jr.

% Computes the unbiased/biased auto correlation sequence (ACS) or cross
% correlation sequence (CCS) estimate. Most computationally efficient if
% data vector length is a power of 2.  Returns length 2*max_lag + 1
% correlation sequence in a column vector.  Zero lag of correlation falls
% in middle of the sequence at element (row) max_lag + 1.  'biased'
% estimates scaled by 1/length(x); 'unbiased' estimates scaled by
% 1/[length(x) - lag index].
%
%     ACS:  r = correlation_sequence(max_lag,bias,x)
%     CCS:  r = correlation_sequence(max_lag,bias,x,y)
%
% max_lag -- maximum time index for estimated ACS or CCS
% bias    -- select: 'unbiased' or 'biased' estimates
% x       -- vector of data samples from channel x
% y       -- vector of data samples from channel y (CCS only)
% r       -- ACS or CCS vector ordered from -max_lag to +max_lag;
%            zero lag is element max_lag + 1

nx = length(x);
if max_lag >= nx, error(' Parameter MAX_LAG > length(x).'), end
```

```
if nargin < 4
    temp = fft([x(:);zeros(nx,1)]);
    r = fftshift(ifft(real(temp).^2 + imag(temp).^2));
else
    ny = length(y);
    if ny ~= nx, error(' Data vectors x and y not of equal length.'), end
    r = fftshift(ifft(fft([x(:);zeros(nx,1)]).*conj(fft([y(:);zeros(nx,1)]))));
end
r = r(nx+1-max_lag:nx+1+max_lag);            % truncate to requested max_lag
if strcmp(bias,'biased')
    r = r/nx;
else                                         % default is 'unbiased'
    r = r./(nx-abs(max_lag-(0:2*max_lag)))';
end

if all(all(imag(x)==0))    % test if x signal is real-valued only
    r = real(r);   % deletes any residual imag part due to FFT calculation
end
```

correlation_sequence_MC

Assuming a multichannel signal record $\mathbf{x}[n]$ of M channels at time index n for which $1 \leq n \leq N$ (see *Digital Spectral Analysis* Section 15.4), the multichannel $M \times M$ correlation matrix at lag index k has the structure

$$\mathbf{r}_{\mathbf{xx}}[k] = \begin{pmatrix} r_{11}[k] & r_{12}[k] & \dots & r_{1M}[k] \\ r_{21}[k] & r_{22}[k] & \dots & r_{2M}[k] \\ \vdots & \vdots & \ddots & \vdots \\ r_{M1}[k] & r_{M2}[k] & \dots & r_{MM}[k] \end{pmatrix}.$$

The estimated autocorrelation and cross correlation elements of this multichannel correlation matrix are obtained from function `correlation_sequence`. The signal record to this function is $\mathbf{x}(n, m)$ for which $n = 1, \dots, N$ and $m = 1, \dots, M$. The correlation matrices are stacked as a block column vector from `-max_lag` at the top of the block column vector to `+max_lag` at the bottom of the block column vector. This means r has scalar dimension M by $M(2\,\texttt{max\_lag} + 1)$. See `correlation_sequence` for discussion of the bias parameter.

```
function r=correlation_sequence_MC(max_lag,bias,x)
% Copyright (c) 2019 by S. Lawrence Marple Jr.

% Generates the multichannel matrix correlation sequence from data.
%
%         r = correlation_sequence_MC(max_lag,bias,x)
%
% max_lag -- maximum correlation lag time index
% bias    -- selection: 'unbiased' or 'biased'
% x       -- sample data array:   x(sample #,channel #)
% r       -- block column vector correlation sequence

[num_pts,num_chs] = size(x);
k = 0:num_chs:max_lag*num_chs;

for m=1:num_chs
    correlation = correlation_sequence(max_lag,bias,x(:,m));
    r(k+m,m) = correlation(max_lag+1:2*max_lag+1);
    for n=m+1:num_chs
        correlation = correlation_sequence(max_lag,bias,x(:,m),x(:,n));
        r(k+m,n) = correlation(max_lag+1:2*max_lag+1);
        r(k+n,m) = conj(correlation(max_lag+1:-1:1));
    end
end
```

correlation_sequence_2D

The biased 2-D autocorrelation estimate at 2-D lag $[k,l]$ is generated from the 2-D signal array x (dimension M in row direction and dimension N in the column direction)

$$r_{xx}[k,l] = \begin{cases} \dfrac{1}{MN} \displaystyle\sum_{m=0}^{M-1-k} \sum_{n=0}^{N-1-l} x[m+k,n+l]x^*[m,n] & \text{for } k \geq 0, l \geq 0 \\ \dfrac{1}{MN} \displaystyle\sum_{m=0}^{M-1-k} \sum_{n=-l}^{N-1} x[m+k,n+l]x^*[m,n] & \text{for } k \geq 0, l < 0 \\ r_{xx}^*[-k,-l] & \text{for } k \leq 0, \text{any } l \end{cases}$$

over a lag range of $|k| \leq$ **max_lag1** and $|l| \leq$ **max_lag2**. The maximum possible lag values are limited to $M-1$ and $N-1$. The desired maximum lag values are determined by the assigned **max_lag1** in the row direction and **max_lag2** in the column direction. The unbiased 2-D ACS $r_{xx}[k,l]$ may by computed by simply substituting the divisor $(M-k)(N-l)$ for MN. See *Digital Spectral Analysis* Section 16.5 for more details. The scalar dimension of the calculated r is (**max_lag1**+1) by (2***max_lag2**+1). Note the relationship cited in the function narrative between MATLAB index value of array r and the lag values.

```
function r=correlation_sequence_2D(max_lag1,max_lag2,x)
% Copyright (c) 2019 by S. Lawrence Marple Jr.

% Estimate the two-dimensional (2-D) autocorrelation sequence (ACS) based on
% the biased estimator of equation (16.34). The computation is most efficient
% if the row and column dimensions of the data array are powers of 2.  A 2-D
% ACS array r of dimension (max_lag1+1) x (2*max_lag2+1) is returned.  The (0,0
% time lag index falls in the middle of the top row of array at element
% (1,max_lag2+1).
%
%     r = correlation_sequence_2D(max_lag1,max_lag2,x)
%
% max_lag1 -- maximum 2-D autocorrelation sequence (ACS) lag index along rows
% max_lag2 -- maximum 2-D autocorrelation sequence lag index along columns
% x        -- two-dimensional data array:  x(row sample #,column sample #)
% r        -- two-dimensional ACS array of dimension  (max_lag1+1) rows x
%                 (2*max_lag2+1) columns
%
% Elements of r are stored as follows:
%
%           ACS Lag Index (row,col)           Actual Array Index (row,col)
%    (0,-max_lag2) . . . . (0,max_lag2)        (1,1). . . . (1,2*max_lag2+1)
%          .                   .        -->     .                .
%          .                   .                .                .
%(max_lag1,-max_lag2).(max_lag1,max_lag2)(max_lag1+1,1).(max_lag1+1,2*max_lag2+1)

[m,n] = size(x);
```

```
if max_lag1 >= m
    error(' Row lag index max_lag1 exceeds available data.')
end
if max_lag2 >= n
    error(' Column lag index max_lag2 exceeds available data.')
end
temp = fft2(x,2*m,2*n);
r = fftshift(ifft2(real(temp).^2 + imag(temp).^2));
r = r(m+1:m+max_lag1+1,n-max_lag2+1:n+max_lag2+1)/(m*n); % truncate to lag size
```

correlogram_psd

This function generates the correlogram-based (either autocorrelation or cross correlation) power spectral density (psd) estimates using the method pioneered by Blackman and Tukey. See *Digital Spectral Analysis* Section 5.6 for details.

$$\mathtt{psd}_{\mathrm{BT}}(f_k) = T \sum_{m=-\mathtt{max\_lag}}^{\mathtt{max\_lag}} \mathtt{window}[m] r[m] \exp(-j2\pi f_k m T).$$

If using the unbiased autocorrelation estimate $r_{xx}[m]$, the auto spectra is computed. If using the unbiased cross correlation estimate $r_{xy}[m]$ for two signal records, the cross spectra is computed. T is the sampling interval in seconds. One of three built-in windows may be selected. No window produces the highest resolution, but also has the highest sidelobe artifacts. A Hamming window suppresses sidelobe artifacts down about 45 dB. A Nuttall window suppresses sidelobes by over 60 dB, but has the lowest resolution of the three windows. The resulting psd is a row vector of num_psd values. The frequency f_k in Hz associated with the k indexing is $f_k = (k - \mathtt{num\_psd}/2)/(\mathtt{num\_psd} * T)$ for $k = 1, \ldots, \mathtt{num\_psd}$.

```
function psd=correlogram_psd(num_psd,window,max_lag,T,x,y)
% Copyright (c) 2019 by S. Lawrence Marple Jr.

% Classical Blackman-Tukey auto/cross spectral estimates based on the Fourier
% transform of the auto/cross correlogram ( ACS  or  CCS  estimate).
%
%      Auto:   psd = correlogram_psd(num_psd,window,max_lag,T,x)
%      Cross:  psd = correlogram_psd(num_psd,window,max_lag,t,x,y)
%
% num_psd -- number of psd vector elements (must be power of 2); frequency
%            spacing between elements is F = 1/(num_psd*T) Hz, with psd(1)
%            corresponding to frequency  -1/(2*T) Hz, psd(num_psd/2 + 1)
%            corresponding to  0  Hz, and psd(num_psd) corresponding to
%            1/(2*T) - F  Hz.
% window  -- window selection:  0 -- none, 1 -- Hamming, 2 -- Nuttall
% max_lag -- maximum time lag (in samples) of ACS or CCS estimate
% T       -- sample interval in seconds
% x       -- vector of data samples
% y       -- vector of data samples (cross PSD only); must have length(y)
%            equal length(x)
% psd     -- vector of num_psd auto/cross PSD values

if max_lag >= length(x), error(' Selected maximum lag exceeds data size.'), end

ph = pi*(-max_lag:max_lag)/max_lag;
if      window == 0
    w = ones(2*max_lag+1,1);          % rectangular (aka boxcar)
elseif window == 1
    w = (.53836 + .46164*cos(ph))';                      % Hamming
elseif window == 2
    w = (.42323 + .49755*cos(ph) + .07922*cos(2*ph))'; % Nuttall
else
```

```
    error(' Window selection number is invalid.')
end
% see MATLAB Signal Processing Toolbox for other window choices

if nargin < 6
    r = w.*correlation_sequence(max_lag,'unbiased',x);
else
    r = w.*correlation_sequence(max_lag,'unbiased',x,y);
end
s = r(max_lag + 1:2*max_lag + 1);
s(num_psd - max_lag + 1:num_psd) = r(1:max_lag);
psd = fftshift(T*fft(s));
```

correlogram_psd_MC

This function computes the multichannel version of the classical Blackman-Tukey correlogram-based auto and cross power spectral density estimator (see *Digital Spectral Analysis* Section 15.5). A multichannel signal matrix $\mathbf{x}(n,m)$ of dimension N row samples by M column signal channels is provided as an input from which the dimension $M \times M$ matrix

$$\mathbf{psd}(f_k) == \begin{pmatrix} \mathrm{psd}_{11}(f_k) & \mathrm{psd}_{12}(f_k) & \dots & \mathrm{psd}_{1M}(f_k) \\ \mathrm{psd}_{21}(f_k) & \mathrm{psd}_{22}(f_k) & \dots & \mathrm{psd}_{2M}(f_k) \\ \vdots & \vdots & \ddots & \vdots \\ \mathrm{psd}_{M1}(f_k) & \mathrm{psd}_{M2}(f_k) & \dots & \mathrm{psd}_{MM}(f_k) \end{pmatrix}$$

is formed at each frequency f_k by populating the matrix with auto and cross psd values generated by auto and cross psd estimates provided by `correlogram_psd`. Each individual frequency **psd** matrix are stacked to form a block column vector of `num_psd` elements. The corresponding scalar dimension of this block column vector is $M *$ `num_psd` by M. The frequency f_k in Hz associated with the **psd** matrix k indexing is $f_k = (k - \texttt{num\_psd}/2)/(\texttt{num\_psd} * T)$ for $k = 1, \dots, \texttt{num\_psd}$. See function `correlogram_psd` for a description of the function input parameters `window`, `max_lag` and T.

```
function psd=correlogram_psd_MC(num_psd,window,max_lag,T,x)
% Copyright (c) 2019 by S. Lawrence Marple Jr.

% This function produces the classical multichannel Blackman-Tukey power
% spectral density matrix over 'num_psd' frequencies.
%
%           psd = correlogram_psd_MC(num_psd,window,max_lag,T,x)
%
% num_psd -- number of frequency points in psd (power of 2)
% window  -- selection:  0 -- none , 1 -- Hamming , 2 -- Nuttall
% max_lag -- maximum lag for correlation matrix estimate
% T       -- sample interval in seconds
% x       -- matrix of data samples:  x(sample #,channel #)
% psd     -- block vector of 'num_psd' psd matrices

[num_pts,num_chs] = size(x);
k = 0:num_chs:(num_psd - 1)*num_chs;

for m=1:num_chs
    psd(k+m,m) = real(correlogram_psd(num_psd,window,max_lag,T,x(:,m)));
    for n=m+1:num_chs
        power = correlogram_psd(num_psd,window,max_lag,T,x(:,m),x(:,n));
        psd(k+m,n) = power;
        psd(k+n,m) = conj(power);
    end
end
```

correlogram_psd_2D

This function computes the two-dimensional (2D) version of the classical Blackman-Tukey correlogram-based power spectral density estimator (see *Digital Spectral Analysis* Section 16.5). A 2-D signal matrix $\mathrm{x}(n1, n2)$ of dimension $N1$ by $N2$ samples is provided as an input from which the dimension ($\texttt{max\_lag1} + 1$) by ($\texttt{max\_lag2} + 1$) 2-D biased (effectively a triangular weighting) autocorrelation matrix **r** is estimated using function `correlation_sequence_2D`. From these, the 2-D correlogram-based `psd` power spectral density is formed from

$$\texttt{psd}(f1_k, f2_l) = T1 * T2 \sum_{k=-\texttt{max\_lag1}}^{\texttt{max\_lag1}} \sum_{l=-\texttt{max\_lag2}}^{\texttt{max\_lag2}} \mathrm{r}[k,l] \exp(-j2\pi[f1 * k * T1 + f2 * l * T2]).$$

The function output `psd` has dimension `num_psd1` by `num_psd2`. The row frequency indexing of the 2-D `psd` matrix is $f1_k = (k - \texttt{num\_psd1}/2)/(\texttt{num\_psd1} * T_1)$ for $k = 1, \ldots, \texttt{num\_psd1}$. The column frequency indexing $f2_l = (l - \texttt{num\_psd2}/2)/(\texttt{num\_psd2} * T_2)$ for $l = 1, \ldots, \texttt{num\_psd2}$.

```
function psd=correlogram_psd_2D(num_psd1,num_psd2,max_lag1,max_lag2,T1,T2,x)
% Copyright (c) 2019 by S. Lawrence Marple Jr.

% Two-dimensional(2-D) classical Blackman-Tukey classical autospectral estimator
% based on the 2-D Fourier transform of the 2-D autocorrelation array estimate.
%
%     psd = correlogram_psd_2D(num_psd1,num_psd2,max_lag1,max_lag2,T1,T2,x)
%
% num_psd1 -- number of row PSD values to evaluate (must be a power of two)
% num_psd2 -- number of column PSD values to evaluate (must be a power of two)
% max_lag1 -- maximum row lag index of 2-D autocorrelation sequence
% max_lag2 -- maximum column lag index of 2-D autocorrelaiton sequence
% T1       -- sample interval along rows in seconds
% T2       -- sample interval along columns in seconds
% x        -- two dimensional sample data array
% psd      -- two dimensional power spectral density (PSD) array

% future window would go here
r = correlation_sequence_2D(max_lag1,max_lag2,x);

s(1:max_lag1+1,1:max_lag2+1) = r(1:max_lag1+1,max_lag2+1:2*max_lag2+1);
s(1:max_lag1+1,num_psd2-max_lag2+1:num_psd2) = r(1:max_lag1+1,1:max_lag2);
s(num_psd1:-1:num_psd1-max_lag1+1,num_psd2:-1:num_psd2-max_lag2+1) = ...
    conj(r(2:max_lag1+1,max_lag2+2:2*max_lag2+1));
s(num_psd1:-1:num_psd1-max_lag1+1,1:max_lag2+1) = ...
    conj(r(2:max_lag1+1,max_lag2+1:-1:1));

psd = fftshift(T1*T2*real(fft2(s)));
```

covariance_lp

This function solves the matrix equation for the minimization of the forward linear prediction (LP) squared error over a finite data record of N samples and order p. As shown in *Digital Spectral Analysis* Section 8.5.1, the solution is given by the $(p+1) \times (p+1)$ set of linear equations for the p linear prediction parameters a^f and the scalar minimum (LP) squared error ρ^f (error variance)

$$\mathbf{T}_p^H \mathbf{T}_p \begin{pmatrix} 1 \\ \mathbf{a}_p^f \end{pmatrix} = \begin{pmatrix} \rho_p^f \\ \mathbf{0}_p \end{pmatrix}.$$

in which

$$\mathbf{T}_p = \begin{pmatrix} x[p+1] & \dots & x[1] \\ \vdots & \ddots & \vdots \\ x[N-p] & & x[p+1] \\ \vdots & \ddots & \vdots \\ x[N] & \dots & x[N-p] \end{pmatrix}$$

is a rectangular Toeplitz matrix of data samples. Normally, solving this matrix equation would require a number of computations proportional to p^3. However, the structure involving the product of Toeplitz data matrices can be exploited to develop a fast solution algorithm with computation proportional to p^2 (see *Digital Spectral Analysis* Section 8.10), which is implemented in this function. Conveniently, the solution of the minimized backward LP squared error ρ^b and the associated backward LP parameters a^b are obtained as well with the same fast algorithm with no additional computational cost. Vectors x, a^f and a^b are all column vectors.

If the LP squared errors are white (or assumed to be), the LP parameters and squared error may be interpreted as AR parameters and the white noise variance, respectively.

```
function [rho_f,a_f,rho_b,a_b]=covariance_lp(p,x)
% Copyright (c) 2019 by S. Lawrence Marple Jr.

% Covariance least squares autoregressive parameter estimation algorithm using
% a fast QR-decomposition type of linear prediction computational solution.
%
%     [rho_f,a_f,rho_b,a_b] = covariance_lp(p,x)
%
% p     -- order of linear prediction/autoregressive filter
% x     -- vector of data samples
% rho_f -- least squares estimate of forward linear prediction variance
% a_f   -- vector of forward linear prediction/autoregressive parameters
% rho_b -- least squares estimate of backward linear prediction variance
% a_b   -- vector of backward linear prediction/autoregressive parameters

%****************  Initialization  ****************

n = length(x);
```

```
if  2*p+1 > n, error(' Order too high; will make solution singular.'), end
a_f = [ ];
a_b = [ ];
r1 = cabs2(x(2:n-1));
r2 = cabs2(x(1));
r3 = cabs2(x(n));
if p <= 0
    rho_f = (r1 + r2 + r3)/n;
    rho_b = rho_f;
    return
end
rho_f = r1 + r3;
rho_b = r1 + r2;
r1 = 1/(rho_b + r3);
c = conj(x(n))*r1;
d = conj(x(1))*r1;
ef = x;
eb = x;
ec = c*x;
ed = d*x;

%*************************  Main Recursion  ************************

for k=1:p

    if (rho_f <= 0) || (rho_b <= 0)
       error(' A prediction squared error was less than or equal to zero.')
    end
    gam = 1 - real(ec(n-k+1));
    del = 1 - real(ed(1));
    if (gam <= 0) || (gam > 1) || (del <= 0) || (del > 1)
       error(' GAM or DEL gain factor not in expected range 0 to 1.')
    end

    % computation for k-th order reflection coefficients
    [eff,ef_k] = splitoff(ef,'top');
    [ebb,eb_n] = splitoff(eb,'bottom');
    delta = ebb'*eff;
    k_f = -delta/rho_b;
    k_b = -conj(delta)/rho_f;

    % order updates for squared prediction errors  rho_f and rho_b
    rho_f = rho_f*(1 - real(k_f*k_b));
    rho_b = rho_b*(1 - real(k_f*k_b));

    % order updates for linear prediction parameter arrays  a_f and a_b
    temp = a_f;
    a_f = [a_f; 0] + k_f*[flipud(a_b);  1];
    a_b = [a_b; 0] + k_b*[flipud(temp); 1];

    % check if maximum order has been reached
    if k == p
        rho_f = rho_f/(n-p);
        rho_b = rho_b/(n-p);
        return
    end
```

```
    % order updates for prediction error arrays  ef and eb
    eb = ebb + k_b*eff;
    ef = eff + k_f*ebb;

    % coefficients for next set of updates
    c1 = ec(1);
    c2 = c1/del;
    c3 = conj(c1)/gam;

    % time updates for gain arrays  c' and d"
    temp = c;
    c = c + c2*d;
    d = d + c3*temp;

    % time updates for ec' and ed"
    temp = ec;
    ec = ec + c2*ed;
    ed = ed + c3*temp;
    [ecc,ec_k] = splitoff(ec,'top');
    [edd,ed_n] = splitoff(ed,'bottom');

    if (rho_f <= 0) || (rho_b <= 0)
       error(' A prediction squared error was less than or equal to zero.')
    end
    gam = 1 - real(ecc(n-k));
    del = 1 - real(edd(1));
    if (gam <= 0) || (gam > 1) || (del <= 0) || (del > 1)
        error(' GAM or DEL gain factor not in expected range 0 to 1.')
    end

    % coefficients for next set of updates
    c1 = ef(1);
    c2 = eb(n-k);
    c3 = conj(c2)/rho_b;
    c4 = conj(c1)/rho_f;
    c5 = c1/del;
    c6 = c2/gam;

    % order updates for c and d; time updates for a_f' and a_b"
    temp = flipud(a_b);
    a_b = a_b + c6*flipud(c);
    c = [c; 0] + c3*[temp; 1];
    temp=a_f;
    a_f = a_f + c5*d;
    d = [0; d] + c4*[1; temp];

    % time updates for rho_f' and rho_b"
    rho_f = rho_f - real(c5*conj(c1));
    rho_b = rho_b - real(c6*conj(c2));

    % order updates for ec and ed; time updates for ef' and eb"
    ec = ecc + c3*eb;
    eb = eb + c6*ecc;
    ed = edd + c4*ef;
    ef = ef + c5*edd;
end
```

covariance_lp_MC

This function implements the extension of `covariance_lp` to the multichannel case for order p. Although not detailed in *Digital Spectral Analysis* Section 15.10.4, the structure of the resulting block matrix equation solution may be exploited to greatly reduce the computation. This fast algorithm has been implemented in the source listing. If the multichannel data vectors of the M channels at time index n are represented as

$$\mathbf{x}[n] = \begin{pmatrix} x_1[n] \\ \vdots \\ x_m[n] \end{pmatrix}$$

[the array x input to this function stores the data as x(sample number n, channel number m)], the minimized multichannel squared error block matrix requires solution of the following pair of multichannel linear prediction block matrix equations

$$\begin{aligned} \underline{\mathbf{a}}_p \underline{\mathbf{R}}_p &= \begin{pmatrix} {\rho_p}^f & \mathbf{0} & \dots & \mathbf{0} \end{pmatrix} \\ \underline{\mathbf{b}}_p \underline{\mathbf{R}}_p &= \begin{pmatrix} \mathbf{0} & \dots & \mathbf{0} & {\rho_p}^b \end{pmatrix}. \end{aligned}$$

The matrix $\underline{\mathbf{R}}_p$ is formed as follows

$$\underline{\mathbf{R}}_p = \begin{pmatrix} \mathbf{R}_{\mathbf{xx}}[0,0] & \dots & \mathbf{R}_{\mathbf{xx}}[0,p] \\ \vdots & & \vdots \\ \mathbf{R}_{\mathbf{xx}}[p,0] & \dots & \mathbf{R}_{\mathbf{xx}}[p,p] \end{pmatrix},$$

which has entries

$$\mathbf{R}_{\mathbf{xx}}[i,j] = \sum_{n=p+1}^{N} \mathbf{x}[n-i]\mathbf{x}^H[n-j].$$

Each block element has dimension $M \times M$, including the forward multichannel LP squared error matrix `rho_a` and backward multichannel LP squared error matrix `rho_b`. The p forward multichannel LP parameter matrices and the p backward multichannel LP parameter matrices are stored as block row vectors. Each therefore has scalar dimension of M by $M * p$. The solution for multichannel LP parameters is also an approximate solution for multichannel AR parameters and associated white noise variance matrix.

```
function [rho_a,a,rho_b,b]=covariance_lp_MC(p,x)

% Multichannel version of the covariance least squares linear prediction
% algorithm using fast computational QR-decomposition-based type of solution.
%
%     [rho_a,a,rho_b,b] = covariance_lp_MC(p,x)
%
% p     -- order of multichannel linear prediction/AR filter
```

```
% x     -- matrix of data samples:  x(sample #,channel #)
% rho_a -- least sqs. estimate of forward linear prediction covariance matrix
% a     -- block row vector of forward linear prediction/AR matrix parameters
% rho_b -- least sqs. estimate of backward linear prediction covariance matrix
% b     -- block row vector of backward linear prediction/AR matrix parameters

%****************  Initialization [ eq (15.xx) ]  ****************

[M,N] = size(x);        % M -- # channels, N -- # samples per channel
if  2*p+1 > N, error('Order too high; will make solution singular.'), end
rho = x*x';
if p <= 0
    rho_a = rho/N;
    rho_b = rho_a;
    return
end
rho_a = rho - x(:,1)*x(:,1)';
rho_b = rho - x(:,N)*x(:,N)';
a = [ ];
b = [ ];
c = x(:,N)'/rho;
d = x(:,1)'/rho;
ea = x;
eb = x;
ec = c*x;
ed = d*x;
I = eye(M,M);
Z = zeros(M,M);
z = zeros(1,M);

%***********************  Main Recursion  ***********************

for k=1:p

    % error condition checks
    if (trace(rho_a) <= 0) || (trace(rho_b) <= 0)
       error('Trace of a covariance matrix was less than or equal to 0.')
    end
    gam = real(ec(N-k+1));
    del = real(ed(1));
    if (gam < 0) || (gam >= 1) || (del < 0) || (del >= 1)
       error('GAM or DEL gain factor not in range 0 to 1.')
    end

    % compute partial correlation and reflection coefficient matrices
    [eaa,ea_k] = breakoff(ea,'left');
    [ebb,eb_N] = breakoff(eb,'right');
    delta = eaa*ebb';
    k_a = -delta/rho_b;
    k_b = -delta'/rho_a;

    % order updates for error covariance matrices   rho_a and rho_b
    rho_a = (I - k_a*k_b)*rho_a;
    rho_b = (I - k_b*k_a)*rho_b;

    % order updates for linear prediction parameter arrays  a and b
```

```
temp = a;
a = [temp Z] + k_a*[b  I];
b = [Z b] + k_b*[I temp];

% check if maximum order has been reached
if k == p
    rho_a = rho_a/(N-p);
    rho_b = rho_b/(N-p);
    return
end

% order updates for prediction error arrays  ea and eb
ea = eaa + k_a*ebb;
eb = ebb + k_b*eaa;

% scalar coefficients for next set of updates
c1 = ec(1)/(1-del);
c2 = ec(1)'/(1-gam);

% time updates for gain vectors  c' and d"
temp = c;
c = temp + c1*d;
d = d + c2*temp;

% time updates of gain "errors"  ec' and ed"
temp = ec;
ec = temp + c1*ed;
ed = ed + c2*temp;
[ecc,ec_k] = breakoff(ec,'left');
[edd,ed_N] = breakoff(ed,'right');

% error condition checks
if (trace(rho_a) <= 0) || (trace(rho_b) <= 0)
   error('Trace of a covariance matrix was less than or equal to 0.')
end
gam = real(ecc(N-k));
del = real(edd(1));
if (gam < 0) || (gam >= 1) || (del < 0) || (del >= 1)
    error('GAM or DEL gain factor not in range 0 to 1.')
end

% vector coefficients for next set of updates
c1 = ea(:,1);
c2 = eb(:,N-k);
c3 = c2'/rho_b;
c4 = c1'/rho_a;
c5 = c1/(1-del);
c6 = c2/(1-gam);

% order updates for c and d; time updates for a' and b"
temp=a;
a = temp + c5*d;
d = [z d] + c4*[I temp];
temp = b;
b = temp + c6*c;
c = [c z] + c3*[temp I];
```

```
    % time updates for rho_a' and rho_b"
    rho_a = rho_a - c5*c1';
    rho_b = rho_b - c6*c2';

    % order updates for ec and ed; time updates for ea' and eb"
    ed = edd + c4*ea;
    ea = ea + c5*edd;
    ec = ecc + c3*eb;
    eb = eb + c6*ecc;
end
```

esd

The energy spectral density (esd) of the complex exponential model h_k and z_k parameters, once estimated, can be used to form an alternative spectral expression to extend the parameter-estimation-only Prony method into a spectral analysis result (see *Digital Spectral Analysis* Section 11.7 for details). The sd is computed using

$$\mathtt{sd}(f_k) = \left|X(f_k)\right|^2$$

over a range of frequencies f_k. There are two possibilities. The one-sided case uses

$$\mathsf{X}_1(z) = T\sum_{k=1}^{p}\left(\frac{h_k}{1 - z_k z^{-1}}\right),$$

and the two-sided case (for slightly sharper peaks) uses

$$\mathsf{X}_2(z) = T\sum_{k=1}^{p} h_k\left(\frac{1}{1 - z_k z^{-1}} - \frac{1}{1 - (z_k^* z)^{-1}}\right).$$

Exponential parameter vectors h and z are column vectors. T is the scalar sampling interval of the data. The sd spectral density is a column vector of num_esd frequency values. The frequency (Hz) ranges over $f_k = (k - \mathtt{num\_esd}/2)/(\mathtt{num\_esd} * T)$ for $k = 1, \ldots, \mathtt{num\_esd}$.

```
function sd=esd(num_esd,method,T,h,z)
% Copyright (c) 2019 by S. Lawrence Marple Jr.

% Computes the one-sided or two-sided energy spectral density
%
%          sd = esd(num_esd,method,T,h,z)
%
% num_esd -- number of ESD values to compute (must be power of two)
% method  -- select ESD model:  1 -- one sided , 2 -- two sided
% T       -- sample interval in seconds
% h       -- vector of complex amplitude parameters
% z       -- vector of complex exponent parameters
% sd      -- vector of energy spectral density values

f = -.5:1/num_esd:.5-1/num_esd;
zi = exp(-1i*2*pi*f);
sd = zeros(num_esd,1);
if  method == 1                    % ESD of one-sided exponential model
    for k=1:num_esd
        sd(k) =  sum(h./(1 - z*zi(k)));
    end
else                              % default:  ESD of two-sided exponential model
    zc = ones(length(z),1)./conj(z);
    for k=1:num_esd
        zip = zi(k);
        sd(k) = sum((zip*h.*(z - zc))./(1 - (z + zc)*zip + (z.*zc)*zip*zip));
    end
end
sd = T*(real(sd).^2 + imag(sd).^2);
```

exponential_parameters

This function converts the complex-valued h and z parameters estimated for an exponential model from signal samples into real-valued damping factor, sinusoidal frequency, amplitude and phase (see *Digital Spectral Analysis* Section 11.4) using

$$\begin{aligned} \alpha_i &= \ln\left|z_i\right|/T \quad \text{sec}^{-1} \\ f_i &= \tan^{-1}\left[\text{Im}\{z_i\}/\text{Re}\{z_i\}\right]/2\pi T \quad \text{Hz} \\ A_i &= \left|h_i\right| \\ \theta_i &= \tan^{-1}\left[\text{Im}\{h_i\}/\text{Re}\{h_i\}\right] \quad \text{radians.} \end{aligned}$$

All parameters into and out of this function are column vectors, except T which is the sampling interval (sec).

```
function [amp,damp,freq,phase]=exponential_parameters(T,h,z)

% Computes the real amplitude, damping factor, frequency, and phase from the
% complex amplitude and exponent factors estimated by the Prony method.
%
%     [amp,damp,freq,phase] = exponential_parameters(T,h,z)
%
% T     -- sample interval in seconds
% h     -- vector of complex amplitude factors
% z     -- vector of complex exponential factors
% amp   -- vector of amplitudes
% damp  -- vector of damping factors (1/sec units)
% freq  -- vector of frequencies (Hz)
% phase -- vector of phases (radians)

freq  = atan2(imag(z),real(z))/(2*pi*T);
damp  = log(abs(z))/T;
amp   = abs(h);
phase = atan2(imag(h),real(h));
```

fast_rls

This function performs the least squares time recursive updating (using a fast algorithm) of forward and backward linear prediction (LP) parameter arrays a^f and a^b (see *Digital Spectral Analysis* Section 9.4). It is called time sequentially one sample at a time. Internal variables related to gain (c and $gamma$) that are updated sequentially are included at input and then again at output once updated. The factor $omega$ controls the rate of adaption, with factors 0.9 to 1 yielding fast adaption. The key sequential update from time index N to $N+1$ is

$$\begin{aligned}\mathbf{a}_{p,N+1} &= \mathbf{a}_{p,N} - \mathbf{P}_N \mathbf{x}^*_{p-1}[N](\mathbf{x}^T_{p-1}[N]\mathbf{a}_{p,N} + x[N+1]) \\ &= \mathbf{a}_{p,N} - e^f_{p,N}[N+1]\mathbf{P}_N \mathbf{x}^*_{p-1}[N] \\ &= \mathbf{a}_{p,N} - e^f_{p,N}[N+1]\mathbf{c}_{p-1,N}.\end{aligned}$$

All parameter arrays for input and output are column vectors.

```
function [init,rho_f,a_f,rho_b,a_b,gamma,c]=...
                      fast_rls(init,omega,x,rho_f,a_f,rho_b,a_b,gamma,c)
% Copyright (c) 2019 by S. Lawrence Marple Jr.

% Fast recursive-least-squares (RLS) sequential adaptive linear prediction/AR
% algorithm with two redundant variable computations to assure long-term
% algorithm numerical stability.
%
%     [init,rho_f,a_f,rho_b,a_b,gamma,c] =
%                     fast_rls(init,omega,x,rho_f,a_f,rho_b,a_b,gamma,c)
%
% init  -- set to 0 to initialize algorithm; function will
%                  auto increment this variable with each iteration
% omega -- exponential weighting factor (between 0 and 1)
% x     -- vector of most recent data samples in linear prediction error filter
% rho_f -- forward linear prediction squared error
% a_f   -- vector of forward linear prediction parameters
% rho_b -- backward linear prediction squared error
% a_b   -- vector of backward linear prediction parameters
% gamma -- gain factor
% c     -- vector of gain parameters

%*************************** Initialization Section ****************************

if init == 0
    a_f = [ ];
    c = [ ];
    gamma = 1;
    rho_f = cabs2(x(1));
    init = init + 1;
    return
elseif init < length(x)
    ef = [1 a_f.']*x(1:init);
    rho_f = omega*rho_f;
    gamma = gamma + cabs2(ef)/rho_f;
    c = [0; c] + (conj(ef)/rho_f)*[1; a_f];
```

```
    temp = x(init+1);
    a_f = [a_f; -ef/temp];
    if init == length(x)-1
        a_b = (-temp/gamma)*flipud(c);
        rho_b = cabs2(temp)/gamma;
    end
    init = init + 1;
    return
end

%******************* Post-Initialization Recursion Section *******************

%adjust1 = 2;                    % redundant gain adjustment factor for EB
%adjust2 = 2;                    % redundant gain adjustment factor for GAMMA

% Forward prediction component update subsection

temp = omega*rho_f;
ef = [1 a_f.']*x;
epf = ef/gamma;
rho_f = temp + real(epf*conj(ef));
if  rho_f <= 0
    error([' rho_f was negative: ',num2str(rho_f)])
end
temp = conj(ef)/temp;
gamma = gamma + real(ef*temp);
if (1/gamma<=0)||(1/gamma>1)
    error([' gamma_f wrong: ',num2str(gamma)])
end
cc = c;
c = [0; cc] + temp*[1; a_f];
a_f = a_f - epf*cc;

% Backward prediction component update subsection

temp = omega*rho_b;
[cc,cp] = splitoff(c,'bottom');
eb = temp*conj(cp);
%eb_redundant = [1 a_b.']*flipud(x);
%eb = eb + adjust1*(eb_redundant - eb);
gamma = gamma - real(eb*cp);
if (1/gamma<=0)||(1/gamma>1)
    error([' gamma_b wrong: ',num2str(gamma)])
end
epb = eb/gamma;
rho_b = temp + real(epb*conj(eb));
if  rho_b <= 0
    error([' rho_b was negative: ',num2str(rho_f)])
end
%gamma_redundant = rho_f/(rho_b*omega^(length(x)-1));
%gamma = gamma + adjust2*(gamma_redundant - gamma);  % stabilization
c = cc - cp*flipud(a_b);
a_b = a_b - epb*flipud(c);
```

hermtoep_lineqs

This function solves a set of $M+1$ linear equations (complex-valued or real-valued) involving a Hermitian Toeplitz matrix (see *Digital Spectral Analysis* Section 3.8.3)

$$\mathbf{H}_M \mathbf{x}_M = \mathbf{z}_M,$$

in which $\mathbf{z}_M$ is a known right-hand side $(M+1)$-dimensional column vector, and

$$\mathbf{H}_M = \begin{pmatrix} t[0] & t^*[1] & \cdots & t^*[M] \\ t[1] & t[0] & \cdots & t^*[M-1] \\ \vdots & \vdots & \ddots & \vdots \\ t[M] & t[M-1] & \cdots & t[0] \end{pmatrix}$$

is the Hermitian Toeplitz matrix. The matrix is not explicitly needed as the left-hand side column vector contains all the elements needed to form $\mathbf{H}$. Exploitation of the structure of the matrix results in a fast algorithm requiring computation proportional to p^2 rather than the normal p^3 computations for linear equations solution.

```
function x=hermtoep_lineqs(t,z)
% Copyright (c) 2019 by S. Lawrence Marple Jr.

% Finds solution of Hermitian Toeplitz set of linear equations  Tx = z .
%
%     x = hermtoep_lineqs(t,z)
%
% t -- left-side column vector of Toeplitz matrix consisting of elements
%      t(0:m) [ (1:m+1) in MATLAB ]
% z -- right-hand-side vector with elements z(0:m) [ (1:m+1) in MATLAB ]
% x -- solution vector with elements x(0:m)       [ (1:m+1) in MATLAB ]

%*****************  Initialization  *****************

a = [ ];
rho = t(1);
if rho == 0, error(' Element t(1) cannot be zero.'), end
x = z(1)/rho;

%*****************  Main Recursion  *****************

for k = 2:length(t)
    kp = -[1 a.']*flipud(t(2:k))/rho;
    rho = rho*(1-cabs2(kp));
    if rho <= 0, error(' Ill-condition: rho <= 0 .'), end
    alpha = (z(k) - x.'*flipud(t(2:k)))/rho;
    a = [a;0] + kp*[flipud(conj(a));1];
    x = [x;0] + alpha*[flipud(conj(a));1];
end
```

lattice

This function modifies the fast algorithm of `levinson_recursion` to work directly with signal samples rather than autocorrelation estimates, in order to achieve improved autoregressive (AR) parameter estimates (and, by implication, improved AR spectral density estimates). A least squares procedure is applied to the signal samples, or the forward and backward linear prediction (LP) errors derived from the samples, to estimate the lattice filter parameters k_p at model order p (see *Digital Spectral Analysis* Section 8.4), also called reflection coefficients. This replaces the step in `levinson_recursion` that estimates the lattice parameter k_p from the autocorrelation for lags 0 to p. Once estimated, the AR parameters are obtained recursively in order using

$$a_p[n] = a_{p-1}[n] + \mathsf{k}_p a^*_{p-1}[p-n]$$

for $n = 1$ to $p-1$. Two variants of the least squares procedure for estimating the lattice parameter k_p have been devised: the harmonic form (by John Burg) and the geometric form. Both are available in this function by setting input parameter `method` to either string `geometric` or `burg`. The estimated AR parameters are in a column vector a of dimension p.

```
function [rho,a]=lattice(method,p,x)
% Copyright (c) 2019 by S. Lawrence Marple Jr.

% Lattice autoregressive parameter estimation algorithms based on the geometric
% or Burg algorithms that estimate the reflection coefficients, then apply
% the Levinson algorithm to get the AR parameters from the reflec. coeffs.
%
%     [rho,a] = lattice(method,p,x)
%
% method -- selection choices: 'geometric' , 'burg'
% p      -- order of autoregressive (AR) process
% x      -- vector of data samples
% rho    -- noise variance estimate
% a      -- vector of autoregressive parameters

%************************  Initialization  ***********************

n = length(x);
a = [ ];
rho = cabs2(x);
den = 2*rho;
rho = rho/n;
temp = 1;
ef = x;
eb = x;

%************************  Main Recursion  ************************

for k = 1:p
    [eff,ef_k] = splitoff(ef,'top');
```

```
    [ebb,eb_n] = splitoff(eb,'bottom');
    if strcmp(method,'geometric')
        k_p = -ebb'*eff/(sqrt(cabs2(ebb))*sqrt(cabs2(eff)));
    elseif strcmp(method,'burg')
        den = den*temp - cabs2(ef_k) - cabs2(eb_n);
        k_p = -2*ebb'*eff/den;
    else
        error('No such lattice AR method.')
    end
    temp = 1 - cabs2(k_p);
    a = [a; 0] + k_p*[flipud(conj(a)); 1];
    rho = rho*temp;
    if rho <= 0
       error('Numerical ill conditioning: rho <= 0')
    end
    if  k == p , return , end
    ef = eff + k_p*ebb;
    eb = ebb + conj(k_p)*eff;
end
```

lattice_MC

This function implements the extension of `lattice` for a single channel signal to the multichannel signal case for order p and M channels. As detailed in *Digital Spectral Analysis* Sections 15.8, 15.10.2, and 15.10.3, the structure of the resulting block Toeplitz matrix equation solution may be exploited to greatly reduce the computations. This fast algorithm has been implemented in the source listing.

Unlike `lattice` that involves a single scalar reflection coefficient for both forward and backward linear prediction that is estimated by a least squares procedure involving a single signal, the multichannel case involves separate $M \times M$ matrix reflection coefficients for the forward and backward linear prediction cases. The extension of the single channel geometric procedure for estimating the matrix reflection coefficients was developed by Vieira and Morf (Section 15.10.2). The extension of the single channel harmonic procedure for estimating the matrix reflection coefficients was developed by Nuttall and Strand (Section 15.10.3). The key equations in the multichannel lattice-based linear prediction algorithm are

$$\begin{aligned} \underline{\mathbf{a}}_{p+1} &= \begin{pmatrix} \underline{\mathbf{a}}_p & \mathbf{0} \end{pmatrix} + \mathbf{A}_{p+1}[p+1] \begin{pmatrix} \mathbf{0} & \underline{\mathbf{b}}_p \end{pmatrix} \\ \underline{\mathbf{b}}_{p+1} &= \begin{pmatrix} \mathbf{0} & \underline{\mathbf{b}}_p \end{pmatrix} + \mathbf{B}_{p+1}[p+1] \begin{pmatrix} \underline{\mathbf{a}}_p & \mathbf{0} \end{pmatrix} \end{aligned}$$

in which the matrix reflection coefficients are incorporated through the relationships

$$\begin{aligned} \mathbf{A}_{p+1}[p+1] &= -\rho^a_{p+1} \left(\mathbf{P}_p{}^b\right)^{-1} \\ \mathbf{B}_{p+1}[p+1] &= -\rho^b_{p+1} \left(\mathbf{P}_p{}^f\right)^{-1}. \end{aligned}$$

Each block element of a and b has dimension $M \times M$, as well as the forward multichannel LP squared error matrix ρ^a and backward multichannel LP squared error matrix ρ^b. The p forward multichannel LP parameter matrices and the p backward multichannel LP parameter matrices are stored as block row vectors. Each therefore has scalar dimension of M by $M * p$. The solution for multichannel LP parameters is also an approximate solution for multichannel AR parameters and associated white noise variance matrix.

Note that if $M = 1$, the single channel case, this function will produce identical numerical results as the geometric and Burg (harmonic) cases computed by `lattice`.

```
function [rho_a,a,rho_b,b]=lattice_MC(method,p,x)
% Copyright (c) 2019 by S. Lawrence Marple Jr.

% Computes multichannel autoregressive parameter matrix using either the
% Vieira-Morf algorithm (multichannel generalization of the single-channel
% geometric lattice algorithm) or the Nuttall-Strand algorithm (multichannel
% generalization of the single-channel Burg lattice algorithm).
```

```
%
%          [rho_a,a,rho_b,b] = lattice_MC(method,p,x)
%
% method -- select: 'VM' -- Vieira-Morf , 'NS' -- Nuttall-Strand
% p      -- order of multichannel AR filter
% x      -- sample data array:  x(sample #,channel #)
% rho_a  -- forward linear prediction error/white noise covariance
%           matrix
% a      -- block vector of forward linear prediction/autoregressive
%           matrix elements
% rho_b  -- backward linear prediction error/white noise covariance
%           matrix
% b      -- block vector of backward linear prediction/autoregressive
%           matrix elements

%********************  Initialization  ********************

[num_pts,num_chs] = size(x);
rho_a = (x.')*conj(x)/num_pts;
rho_b = rho_a;
ef = x.';
eb = x.';
I = eye(num_chs,num_chs);
Z = zeros(num_chs,num_chs);
a = [ ];
b = [ ];

%********************  Main Recursion  ********************

for k=1:p

% update estimated error covariance matrices
    n = k+1:num_pts;
    rho_a_hat = ef(:,n)*ef(:,n)';
    rho_b_hat = eb(:,n-1)*eb(:,n-1)';
    rho_ab_hat = ef(:,n)*eb(:,n-1)';
%*********  Geometric Mean Lattice Algorithm (Vieira-Morf)  *********

    if strcmp(method,'VM')
        % estimate normalized partial correlation matrix
        delta=(hermsqrt(rho_a_hat)')\rho_ab_hat/hermsqrt(rho_b_hat);
        % update forward and backward reflection coefficients
        k_a = -(hermsqrt(rho_a)')*delta/(hermsqrt(rho_b)');
        k_b = -(hermsqrt(rho_b)')*delta'/(hermsqrt(rho_a)');

%********  Harmonic Mean Lattice Algorithm (Nuttall-Strand)  *********

    elseif strcmp(method,'NS')
        % estimate partial correlation matrix
        delta=bilineqn(rho_a_hat/rho_a,rho_b\rho_b_hat,2*rho_ab_hat);
        % update forward and backward reflection coefficients
        k_a = -delta/rho_b;
        k_b = -delta'/rho_a;

%*****************  Error in the parameter method  *******************
```

```
    else
        error(' Method does not correspond to available techniques.')
    end
    % update forward and backward error covariances
    rho_a = (I - k_a*k_b)*rho_a;
    rho_b = (I - k_b*k_a)*rho_b;
    % update forward and backward prediction parameters
    temp = a;
    a = [a Z] + k_a*[b I];
    b = [Z b] + k_b*[I temp];
    % update the error residuals
    if k < p
        temp = ef(:,n);
        ef(:,n) = temp + k_a*eb(:,n-1);
        eb(:,n) = eb(:,n-1) + k_b*temp;
    end
end
```

lattice_2D

Given a 2-D signal for which a 2-D linear prediction(LP)/autoregressive(AR) model of order (p_1, p_2) is to be generated, this function is an extension of the `lattice` function to the 2-D case. It generates the estimated first and fourth quarter plane LP/AR parameter matrices of dimension $(p_1+1)\times(p_2+1)$ and their associated scalar white noise variances p_{Q1} and p_{Q4}, for which the indexing ranges are

$$a[m,n] = \begin{cases} a_1[m,n] & 0 \le m \le p_1 (1 \le n \le p_2), 1 \le m \le p_1 (n=0) \\ a_4[m,n] & 0 \le m \le p_1 (-p_2 \le n \le -1), 1 \le m \le p_1 (n=0) \end{cases}.$$

A general discussion of the 2-D lattice method is presented in *Digital Spectral Analysis, Second Edition* Section 16.7.6, but the detailed development of the fast algorithm source listed below is not covered in this edition.

```
function [p_Q1,a_Q1,p_Q4,a_Q4,P,A]=lattice_2D(p1,p2,x)
% Copyright (c) 2019 by S. Lawrence Marple Jr.

% Computes 2-D quarter-plane support AR parameter arrays given 2-D data set
% x  using Marple algorithm to exploit the doubly Toeplitz structure. Only
% Q1 and Q4 quadrant AR parameter arrays are computed; Q2 and Q3 quarter
% plane arrays are simply the complex conjugates, respectively, of the Q4
% and Q1 arrays, ie, a_Q3=conj(a_Q1), p_Q3=p_Q1, a_Q2=conj(a_Q4), p_Q2=p_Q4.
% Most computationally efficient when p2 >= p1, where p1 is the fixed order
% and p2 is the variable order; else switch roles. Reduces to 1-D Burg lattice
% algorithm results if (N1=1,p1=0). Reduces to modified covariance algorithm
% results if (N2=1,p2=0). Must have p1 < N1 and p2 < N2.
%
% NOTE: may need to do both (p1 fixed/p2 variable) and (p2 variable/p2 fixed)
% to get average results since solutions are different
%
%     [p_Q1,a_Q1,p_Q4,a_Q4,P,A] = lattice_2D(p1,p2,x)
%
% p1   -- row order of 2-D autoregressive (AR) quarter-plane filter
% p2   -- column order of 2-D autoregressive quarter-plane filter
% x    -- 2-D sample data array:  x(row sample#,column sample #)  N1 x N2
% p_Q1 -- Q1-quarter-plane white noise variance estimate
% a_Q1 -- two-dimensional Q1-quarter-plane AR parameter matrix estimates
% p_Q4 -- Q4-quarter-plane white noise variance estimate
% a_Q4 -- two-dimensional Q4-quarter-plane AR parameter matrix estimates
% P    -- internal array of A error variance (used in minimum_variance_psd_2D)
% A    -- internal block array of other parameter matrices (        "         )

%****************************  Initialization  *****************************

[N1,N2] = size(x);
P1 = p1 + 1;
P2 = p2 + 1;
NP = N1 - p1;
colsize = P1:-1:1;
rowsize = P1:N1;
ef = zeros(P1,NP*N2);% order 0 dimension of each 'forward error': (p1+1)x(N1-p1)
```

```
for n = 1:N2
    ef(:,(n-1)*NP+1:n*NP) = toeplitz(x(colsize,n),x(rowsize,n));
end
eb = ef;       % order 0 dimension of each 'backward error': (p1+1)x(N1-p1)
P = hermitian(ef*ef');
P = (P + fliplr(flipud(P.')));% both persymmetric (only at order 0) & hermitian
I = eye(P1,P1);

%********* Compute Special MultiChannel-Like delta, P, and A arrays  ***********

for p = 1:p2

    % update estimated error covariance matrices, dimension (p1+1)x(p1+1)
    n = p*NP+1:N2*NP;
    p_f = hermitian(ef(:,n)*ef(:,n)');
    if any(diag(p_f) <= 0)
        error('Error1')
    end
    p_b = hermitian(eb(:,n-NP)*eb(:,n-NP)');
    if any(diag(p_b) <= 0)
        error('Error2')
    end
    p_fb = ef(:,n)*eb(:,n-NP)';

    % persymmetric partial correlation matrix, dimension (p1+1)x(p1+1)
    delta = p_fb + fliplr(flipud(p_fb.'));      % not hermitian

    % generate reflection coefficient matrix, dimension (p1+1)x(p1+1)
    P = hermitian(p_f + fliplr(flipud(conj(p_b))));
    K = -delta/fliplr(flipud(conj(P)));
    KK = fliplr(flipud(conj(K)));

    % update error covariance matrix, dimension (p1+1)x(p1+1)
    P = hermitian(P + K*delta');
    if any(diag(P) <= 0)
        error(' Error3: diagonal element - when + is expected.')
    end

    % update prediction parameter matrices, dimension (p1+1)x(p1+1)p2
    if p == 1
       A = K;
    else
       A = [A+K*(fliplr(flipud(conj(A)))), K];
    end

    % check if maximum order has been reached
    if p==p2, break, end

    % order update of the error residuals
    temp = ef(:,n);
    ef(:,n) = temp + K*eb(:,n-NP);
    eb(:,n) = eb(:,n-NP) + KK*temp;

end
clear temp ef eb K KK
```

```
%*************  Compute Q1 Two-Dimensional (2-D) AR Array  ****************
P = P*(1/(2*NP*(N2-p2)));

p_inv = [1 zeros(1,p1)]/P;
p_Q1 = 1/real(p_inv(1));
a_Q1 = p_Q1*p_inv;
a_Q1 = [a_Q1 a_Q1*A];
a_Q1 = reshape(a_Q1,P1,P2);

%***************  Compute Q4 Two-Dimensional AR Array  *****************

p_inv = [zeros(1,p1) 1]/P;
p_Q4 = 1/real(p_inv(P1));
a_Q4 = p_Q4*p_inv;
a_Q4 = [a_Q4 a_Q4*A];
a_Q4 = flipud(reshape(a_Q4,P1,P2));
```

levinson_recursion

This function solves the Yule-Walker equations for linear prediction / autoregressive (LP/AR) parameters and associated white noise variance ρ at order p, given the autocorrelation sequence from $r[0]$ to $r[p]$, as described in *Digital Spectral Analysis* Section 7.3.2

$$\begin{pmatrix} r[0] & r^*[1] & \dots & r^*[m] \\ r[1] & r[0] & \dots & r^*[m-1] \\ \vdots & \vdots & \ddots & \vdots \\ r[m] & r[m-1] & \dots & r[0] \end{pmatrix} \begin{pmatrix} 1 \\ a[1] \\ \vdots \\ a[m] \end{pmatrix} = \begin{pmatrix} \rho \\ 0 \\ \vdots \\ 0 \end{pmatrix}.$$

The Toeplitz structure of the matrix equation is exploited to develop a fast algorithm (see Section 3.8) that recursively by order solves for the AR parameters in a number of computation proportional to p^2 rather than the usual p^3 for matrix equation solution. Both a and r are column vectors of dimension p and $p+1$, respectively.

```
function [rho,a]=levinson_recursion(r)
% Copyright (c) 2019 by S. Lawrence Marple Jr.

% Levinson recursion algorithm for solving Hermitian Toeplitz linear equation
% Ra = rho. Typically vector  a  contains the LP/AR parameters and  rho  is
% the white noise variance (see Section 7.3.1)
%
%     [rho,a] = levinson_recursion(r)
%
% r    -- left-side column vector of Toeplitz matrix elements r(1:p+1) for
%         MATLAB  [ r(0:p) in text ]
% rho -- top element of right-hand-side all-zeros vector
% a    -- solution vector a(1:p)

%****************************  Initialization  ****************************

a = [ ];
rho = r(1);
if rho == 0, error(' Element r(1) cannot be zero.'), end

%****************************  Main Recursion  ****************************

for i = 2:length(r)
    gamma = -[1 a.']*r(i:-1:2)/rho;
    rho = rho*(1 - cabs2(gamma));
    if rho <= 0, error(' Numerical ill-conditioning: rho <= 0 .'), end
    a = [a; 0] + gamma*[flipud(conj(a)); 1];
end
```

levinson_recursion_MC

This function is the multichannel extension of the classic Levinson recursion algorithm. Unlike the single channel case, there are separate forward and backward linear prediction / autoregressive (LP/AR) block matrix parameters that form the multichannel Yule-Walker equations (see *Digital Spectral Analysis* Sections 15.7 and 15.8)

$$\underline{\mathbf{a}}_p \underline{\mathbf{R}}_p = \begin{pmatrix} \rho_p{}^f & \mathbf{0} & \ldots & \mathbf{0} \end{pmatrix}.$$

$$\underline{\mathbf{b}}_p \underline{\mathbf{R}}_p = \begin{pmatrix} \mathbf{0} & \ldots & \mathbf{0} & \rho_p{}^b \end{pmatrix},$$

for which the multi-channel block Toeplitz matrix is

$$\underline{\mathbf{R}}_p = \begin{pmatrix} \mathbf{R}_{\mathbf{xx}}[0] & \mathbf{R}_{\mathbf{xx}}[1] & \ldots & \mathbf{R}_{\mathbf{xx}}[p] \\ \mathbf{R}_{\mathbf{xx}}[-1] & \mathbf{R}_{\mathbf{xx}}[0] & \ldots & \mathbf{R}_{\mathbf{xx}}[p-1] \\ \vdots & \vdots & \ddots & \vdots \\ \mathbf{R}_{\mathbf{xx}}[-p] & \mathbf{R}_{\mathbf{xx}}[-p+1] & \ldots & \mathbf{R}_{\mathbf{xx}}[0] \end{pmatrix}$$

and the LP/AR forward and backward parameter block row vectors

$$\underline{\mathbf{a}}_p = \begin{pmatrix} \mathbf{I} & \mathbf{A}_p[1] & \ldots & \mathbf{A}_p[p] \end{pmatrix},$$

$$\underline{\mathbf{b}}_p = \begin{pmatrix} \mathbf{B}_p[p] & \ldots & \mathbf{B}_p[1] & \mathbf{I} \end{pmatrix}.$$

The Toeplitz structure of the matrix equations are exploited to develop a fast algorithm that recursively by order solves for the AR parameters in a number of block matrix computations proportional to p^2 rather than the usual p^3 for block matrix equation solution. Each block element of a and b have dimension $M \times M$, as well as the forward multichannel LP squared error matrix ρ^a and backward multichannel LP squared error matrix ρ^b. The p forward multichannel LP parameter matrices and the p backward multichannel LP parameter matrices are stored as block row vectors. Each therefore has scalar dimension of M by $M * p$. The solution for multichannel LP parameters is also an approximate solution for multichannel AR parameters and associated white noise variance matrix.

Note that for $M = 1$, the single channel case, this function will produce identical numerical results as `levinson_recursion`.

```
function [rho_a,a,rho_b,b]=levinson_recursion_MC(r)
% Copyright (c) 2019 by S. Lawrence Marple Jr.

% Solves for the multichannel AR parameter matrices from the multichannel
% correlation sequence by the multichannel Levinson algorithm.
%
%          [rho_a,a,rho_b,b] = levinson_recursion_MC(r)
%
```

```
% r     -- multichannel correlation sequence block column vector
% rho_a -- forward linear prediction error/white noise covariance matrix
% a     -- forward linear prediction/autoregressive (AR) parameter block vector
% rho_b -- backward linear prediction error/white noise covariance matrix
% b     -- backward linear prediction/autoregressive (AR) parameter block vector

%*******************  Initialization  *******************

[n,num_chs] = size(r);
p = n/num_chs - 1;       %  multichannel autoregressive AR process order
I = eye(num_chs,num_chs);
Z = zeros(num_chs,num_chs);
a = [ ];
b = [ ];
index = 1:num_chs;
rho_a = r(index,:);
rho_b = rho_a;
R = [];

%*******************  Main Recursion  *******************

for k=1:p

    % update cross covariance matrix
    index = index + num_chs;
    R = [r(index,:); R];
    delta = [I a]*R;

    % update forward and backward reflection coefficient matrices
    k_a = -delta/rho_b;
    k_b = -delta'/rho_a;

    % update forward and backward error covariance matrices
    rho_a = (I - k_a*k_b)*rho_a;
    rho_b = (I - k_b*k_a)*rho_b;
    % check for numerical ill-conditioning would go here

    % update forward and backward prediction parameter matrices
    temp = a;
    a = [a Z] + k_a*[b I];
    b = [Z b] + k_b*[I temp];

end
```

levinson_recursion_2D

This function is the extension of function `levinson_recursion` to the 2-D case. An overview is provided in *Digital Spectral Analysis* Section 16.7.4. If a 2-D autocorrelation matrix r of dimension p_1+1 by p_2+1 is used as input, the resulting 2-D linear prediction(LP)/autoregressive(AR) model of order (p_1, p_2) is generated, specifically, for the first and fourth quarter plane LP/AR parameter matrices of dimension $(p_1+1) \times (p_2+1)$ and their associated scalar white noise variances p_{Q1} and p_{Q4}, for which the indexing ranges are

$$a[m,n] = \begin{cases} a_1[m,n] & 0 \le m \le p_1 (1 \le n \le p_2), 1 \le m \le p_1 (n=0) \\ a_4[m,n] & 0 \le m \le p_1 (-p_2 \le n \le -1), 1 \le m \le p_1 (n=0) \end{cases}.$$

```
function [p_Q1,a_Q1,p_Q4,a_Q4]=levinson_recursion_2D(r)
% Copyright (c) 2019 by S. Lawrence Marple Jr.

% Compute 2-D quarter plane AR parameter arrays given 2-D known or estimated
% ACS using two-dimensional Levinson algorithm that exploits the double
% Toeplitz structure. Only Q1 and Q4 quadrant AR parameter arrays are computed;
% Q3 and Q2 quarter arrays are simply the complex conjugates, respectively,
% of the Q1 and Q4 arrays.  Most computationally efficient when  p2 > p1 ,
% else switch orders.
%
%     [p_Q1,a_Q1,p_Q4,a_Q4] = levinson_recursion_2D(r)
%
%  r    -- two-dimensional autocorrelation sequence (ACS) array
%  p_Q1 -- Q1-quarter plane white noise variance
%  a_Q1 -- two-dimensional Q1-quarter plane AR parameter matrix
%  p_Q4 -- Q4-quarter plane white noise variance
%  a_Q4 -- two-dimensional Q4-quarter plane AR parameter matrix

%*********************** Initialization **************************

[m,n] = size(r);
p1 = m-1;
p2 = (n-1)/2;
I = eye(p1+1);
Z = zeros(p1+1);
P = toeplitz(r(1:p1+1,p2+1)',r(1:p1+1,p2+1));
A = [ ];
R = [ ];

%************ Compute 2-D block R, P, and A arrays ************

for p=1:p2
    R = [toeplitz(r(1:p1+1,p2+1-p)',r(1:p1+1,p2+1+p)); R];
    delta = persymmetric([I A]*R);
    K = -delta/fliplr(flipud(conj(P)));
    P = hermitian((I-K*fliplr(flipud(conj(K))))*P);
    A = [A Z] + K*[fliplr(flipud(conj(A))) I];
end
```

```
%*************** Compute Q1 two-dimensional AR array ***************

p_inv = [1 zeros(1,p1)]/P;
p_Q1 = 1/real(p_inv(1));
a_Q1 = p_Q1*p_inv;
a_Q1 = [a_Q1 a_Q1*A];
a_Q1 = reshape(a_Q1,p1+1,p2+1);

%*************** compute Q4 two-dimensional AR array ***************

p_inv = [zeros(1,p1) 1]/P;
p_Q4 = 1/real(p_inv(p1+1));
a_Q4 = p_Q4*p_inv;
a_Q4 = [a_Q4 a_Q4*A];
a_Q4 = flipud(reshape(a_Q4,p1+1,p2+1));
```

lms

This function performs the gradient approximation to the least mean square (LMS) time recursive updating of forward linear prediction (LP) parameter column vector a (see *Digital Spectral Analysis* Section 9.3). It is called time sequentially one sample at a time. The factor mu, which must be experimentally determined, controls the rate of adaption. The key sequential update from time index N to $N+1$ is

$$\mathbf{a}_{p,N+1} = \mathbf{a}_{p,N} - e^f_{p,N}[N+1]2\mu\mathbf{x}^*_{p-1}[N].$$

The $p+1$ most current signal samples are retained in column vectors x.

```
function a=lms(mu,x,a)
% Copyright (c) 2019 by S. Lawrence Marple Jr.

% Least-mean-square (LMS) time-recursive adaptive AR algorithm.
%
%      a = lms(mu,x,a)
%
% mu -- adaptive time constant
% x  -- vector of data samples
% a  -- vector of linear prediction/autoregressive (AR) parameters

ef = [1 a.']*x;
a = a - 2*mu*ef*conj(x(2:length(x)));
```

lstsqs_prony

This function performs the three steps of the least squares Prony technique (see *Digital Spectral Analysis* Section 11.5) from input signal samples in column vector x (may be complex or real valued). The three steps are

- Least squares fit to signal data for linear prediction (LP) parameters for damped sinusoidal model (method = 1) or undamped sinusoidal model (method = 2).

- Form polynomial with LP parameters and factor for complex parameters z.

- With estimates of z, compute least squares fit to obtain complex parameters h.

The resulting approximation of the signal data to the complex exponential model is

$$\hat{x}[n] = \sum_{k=1}^{p} h_k z_k^{n-1}.$$

Note that the num_exps input must be an even number as two complex conjugate z and h parameters are needed for real-valued signal data. The column vectors h and z have dimension num_exps.

```
function [h,z]=lstsqs_prony(method,num_exps,x)
% Copyright (c) 2019 by S. Lawrence Marple Jr.

% Solves for the exponential model parameters by the least squares Prony method
% using fast QR computational algorithms.
%
%          [h,z] = ls_prony(method,num_exps,x)
%
% method   -- select: 1 -- regular Prony (damped exponentials) , 2 -- modified
%                     Prony (undamped exponentials)
% num_exps -- number of presumed exponentials (must be even for method = 2)
% x        -- vector of data samples
% h        -- vector of complex amplitudes of exponential time series model
% z        -- vector of complex exponents of exponential time series model

if  num_exps < 1
    error(' Number of exponentials incorrectly selected.')
end

% first step:  determine linear prediction or linear smoothing parameters
if  method == 1
    [rho_f,a_f,rho_b,a_b ] = covariance_lp(num_exps,x);
    p = [1 a_f.'];
```

```
elseif  method == 2
    [rho_s,p] = symcovar(num_exps,x);
else
    error(' Selected method number is not valid.')
end

% second step:  factor polynomial for complex exponent parameters
z = roots(p);

% third step:  estimate complex amplitude parameters
h = lstsqsvdm(x,z);
```

lstsqsvdm

This function estimates the complex-valued parameters h in a complex exponential approximation model of signal data x

$$\hat{x}[n] = \sum_{k=1}^{p} h_k z_k^{n-1}.$$

The resulting matrix expression for the least squares fit (see *Digital Spectral Analysis* Section 11.5) is

$$\left(\mathbf{Z}^H \mathbf{Z}\right) \mathbf{h} = \left(\mathbf{Z}^H \mathbf{x}\right),$$

in which

$$\mathbf{Z} = \begin{pmatrix} 1 & 1 & \dots & 1 \\ z_1 & z_2 & \dots & z_p \\ \vdots & \vdots & & \vdots \\ z_1^{N-1} & z_2^{N-1} & \dots & z_p^{N-1} \end{pmatrix}, \quad \mathbf{h} = \begin{pmatrix} h_1 \\ h_2 \\ \vdots \\ h_p \end{pmatrix}, \quad \mathbf{x} = \begin{pmatrix} x[1] \\ x[2] \\ \vdots \\ x[N] \end{pmatrix}.$$

The source listing is for a fast computational algorithm that exploits the Vandermonde structure of the rectangular matrix $\mathbf{Z}$. The parameters h and z are column vectors of dimension p and x is a column vector of N samples.

```
function h=lstsqsvdm(x,z)
% Copyright (c) 2019 by S. Lawrence Marple Jr.

% Fast computational solution of least squares Vandermonde normal equations
% based on algorithm of Demeure [1989].
%
%          h = lstsqsvdm(x,z)
%
%  x -- vector of complex data samples
%  z -- vector of complex exponent parameters
%  h -- vector of estimated complex amplitude parameters

%*************************  Initialization  *************************

n = length(x);
num_exps = length(z);
h = zeros(num_exps,1);
phi = zeros(num_exps,1);
x1 = zeros(num_exps,1);
x2 = zeros(num_exps,1);
y1 = zeros(num_exps,1);
y2 = zeros(num_exps,1);
beta1 = zeros(n,1);
beta2 = zeros(n,1);
r = zeros(num_exps);
c1 = z(1);
r1 = cabs2(c1);
```

```
if abs(r1 - 1) <= 1000*eps % use built-in MATLAB 'eps' value (machine precision)
    r2 = n;
else
    r2 = (1 - r1^n)/(1 - r1);
end
phi(1) = r2;
phi_sq_inv = 1/r2;
phi_inv = sqrt(phi_sq_inv);
c2 = conj(c1)^n;
c3 = phi_sq_inv;
q = phi_inv;
c4 = phi_inv*x(1);
beta1(1) = c3;
beta2(1) = c2*c3;
for k = 2:n
    q = q*c1;
    c4 = c4 + conj(q)*x(k);
    c3 = c3*c1;
    beta1(k) = c3;
    beta2(k) = c2*c3;
end
h(1) = c4*phi_inv;
if  num_exps <= 1, return, end
c3 = c1/r1;
c4 = c2*c3;
for k = 1:num_exps-1
    c5 = z(k+1);
    r1 = cabs2(c5);
    c6 = c5/r1;
    c7 = conj(c5)^n;
    c8 = c6*c7;
    if  (abs(c5-c1)/abs(c1)) < 1000*eps
        error(' Two  z  terms (complex exponents) were essentially identical.')
    end
    c9 = conj(c5)*c1;
    c5 = (1 - c9^n)/(1 - c9);
    c9 = c5*phi_sq_inv;
    r(1,k) = c9;
    x1(k) = c6 - c3*c9;
    x2(k) = c4*c9 - c8;
    y1(k) = 1 - c9;
    y2(k) = c7 - c2*c9;
end

%************************  Main Recursion  ************************

for k = 2:num_exps
    k1 = k - 1;
    c1 = x1(k1);
    c2 = x2(k1);
    c3 = conj(y1(k1));
    c4 = conj(y2(k1));
    c5 = z(k);
    r1 = cabs2(c5);
    if  abs(r1 - 1) > 1000*eps
        r2 = real((c1*c3 + c2*c4)/(c5/r1 - c5));
```

```
    else
        r2 = n - (real(r(1:k1,k1)).^2 + imag(r(1:k1,k1)).^2)'*phi(1:k1);
    end
    phi(k) = r2;
    if r2 <= 0
        error(' Negative ''sig'' vector element value encountered.')
    end
    phi_sq_inv = 1/r2;
    phi_inv = sqrt(phi_sq_inv);
    c6 = phi_inv*conj(c3);
    c7 = phi_inv*conj(c4);
    q = -phi_inv*c5*(conj(c1)*(beta1(1) - 1) + conj(c2)*beta2(1));
    c8 = conj(q)*x(1);
    beta1(1) = beta1(1) + c6*q;
    beta2(1) = beta2(1) + c7*q;
    for j = 2:n
        q = c5*(q-phi_inv*(conj(c1)*beta1(j)+conj(c2)*beta2(j)));
        c8 = c8 + conj(q)*x(j);
        beta1(j) = beta1(j) + c6*q;
        beta2(j) = beta2(j) + c7*q;
    end
    h(k) = c8*phi_inv;
    if k < num_exps
        c6 = c5/r1;
        k2 = k:num_exps-1;
        temp=phi_sq_inv*(conj(c1)*y1(k2) + conj(c2)*y2(k2))./(conj(c6-z(k2+1)));
        r(k,k2) = conj(temp)';
        x1(k2) = x1(k2) - c1*temp;
        x2(k2) = x2(k2) - c2*temp;
        y1(k2) = y1(k2) - conj(c3)*temp;
        y2(k2) = y2(k2) - conj(c4)*temp;
    end
end
for k = num_exps-1:-1:1
    h(k) = h(k) - conj(r(k,k:num_exps-1))*h(k+1:num_exps);
end
```

ma

This function generates the q parameters of a column vector b of estimated moving average (MA) parameters with associated white noise variance estimate *rho*. The algorithm uses a high order autoregressive (AR) fit to the data, which is replaced using the AR parameters to fit a lower order MA model, as described in *Digital Spectral Analysis* Section 10.3. The user may specify the order of the AR initial model fit, but if the third input argument to the function is missing, it is arbitrarily set to twice the specified MA order.

```
function [rho,b]=ma(q,x,m)
% Copyright (c) 2019 by S. Lawrence Marple Jr.

% Moving average (MA) parameter estimation algorithm based on a high-order AR
% model estimated with the Yule-Walker algorithm.
%
%     [rho,b] = ma(q,x,m)
%
% q   -- order of moving MA filter
% x   -- vector of data samples
% m   -- order of 'high-order' AR filter
% rho -- white noise variance estimate
% b   -- vector of moving average parameters

% default high-order AR length if third argument omitted
if nargin < 3, m = 2*q; end

% fit high-order AR; can use any AR method (e.g.,'lattice','covariance_lp',
% and 'modcovar_lp' algorithms) in lieu of 'yule_walker' algorithm
[rho,a] = yule_walker(m,x);

% determine MA parameters; other AR methods could be used in lieu of
% 'yule_walker' algorithm
[r,b] = yule_walker(q,[1;a]);
```

minimum_eigenvalue

This function finds the minimum eigenvalue and associated eigenvector of a Hermitian Toeplitz matrix. The power iteration method using the inverse of the matrix is applied until convergence to the minimum eigenvalue and eigenvector, as described in *Digital Spectral Analysis* Section 3.8.5. Only the left side column vector of the Hermitian Toeplitz matrix is used as input. This function is used in the Pisarenko sinusoidal eigen procedure (see function `pisarenko`).

```
function [eigval,eigvec]=minimum_eigenvalue(t)
% Copyright (c) 2019 by S. Lawrence Marple Jr.

% Computes the minimum eigenvalue and associated eigenvector for a Hermitian
% Toeplitz matrix T .
%
%     [eigval,eigvec] = minimum_eigenvalue(t)
%
% t      -- left side column vector of matrix T
% eigval -- minimum eigenvalue of the matrix T
% eigvec -- eigenvector associated with minimum eigenvalue

% Initialize eigval, eigvec, and eps (preset to machine precision by MATLAB)
% May have to be changed if loop does not converge.
eigval = 123.456789;
eigval_old = 0;
eigvec = ones(length(t),1);

while abs(eigval-eigval_old) > eps*eigval_old
    eigval_old = eigval;
    e = hermtoep_lineqs(t,eigvec);
    temp = 1/cabs2(e);
    eigval = real(eigvec'*e)*temp;
    eigvec = e*temp;
end
```

minimum_variance_psd

This function determines the minimum variance power spectral density (psd) row vector using the technique described in *Digital Spectral Analysis* Sections 12.4 and 12.5 that depends on use of order p autoregressive (AR) parameter estimates

$$\mathtt{psd}_{\mathrm{MV}}(f_k) = \frac{T}{\sum_{m=-p}^{p} \psi_{\mathrm{MV}}[m] \exp(-j2\pi f_k mT)}$$

in which the $\psi_{\mathrm{MV}}[m]$ coefficients are calculated as a linearly weighted (windowed) correlation of AR parameters

$$\psi_{\mathrm{MV}}[m] = \begin{cases} \frac{1}{\rho_p} \sum_{i=0}^{p-m} (p+1-m-2i) a_p[m+i] a_p^*[i] & \text{for } 0 \le m \le p \\ \psi_{\mathrm{MV}}^*[-m] & \text{for } -1 \ge m \ge -p. \end{cases}$$

Parameter T is the sampling interval (sec). The *psd* row vector has dimension num_psd for which the frequency (Hz) is $f_k = (k - \mathtt{num\_psd}/2)/(\mathtt{num\_psd} * T)$ for $k = 1, \ldots, \mathtt{num\_psd}$.

```
function psd=minimum_variance_psd(num_psd,p,T,x)
% Copyright (c) 2019 by S. Lawrence Marple Jr.

% This function computes the minimum variance spectral estimator
%
%          psd = minimum_variance_psd(num_psd,p,T,x)
%
% num_psd -- number of power spectral density values (must be power of two)
% p       -- minimum variance adaptive filter order (number of taps = p + 1)
% T       -- sample interval in seconds
% x       -- vector of data samples
% psd     -- vector of power spectral density values

% estimate the AR parameters (use any AR estimation technique)
[rho,a] = lattice('burg',p,x);
%[rho,a] = yule_walker(p,x);             % or
%[rho,a] = lattice('geometric',p,x)  % or
a = [1; a];

% correlate the AR parameters
psi = [ ];
p = p + 1;
for k =1:p
    pk = p - k + 1;
    psi = [psi; (a(1:pk)'*((pk:-2:2-pk)'.*a(k:p)))/rho];
end

% evaluate psi parameters with FFT to yield minimum variance PSD vector
psi(num_psd:-1:num_psd-p+2) = conj(psi(2:p));
psd = fftshift(T*ones(num_psd,1)./(real(fft(psi))));
```

minimum_variance_MC

This function is the multichannel extension of the single channel function `minimum_variance_psd`. Although not detailed in *Digital Spectral Analysis* Section 15.13 and assuming M signal channels, this function will produce the $M \times M$ **psd** block matrix $M \times M$ auto- and cross-power spectral density (PSD) matrices per frequency

$$\mathbf{psd}(f_k) = \begin{pmatrix} P_{11}(f_k) & \dots & P_{1M}(f_k) \\ \vdots & \ddots & \vdots \\ P_{M1}(f_k) & \dots & P_{MM}(_kf) \end{pmatrix},$$

at each frequency f_k to form a block column vector for `num_psd` frequencies. The scalar dimension of the computed `psd` block column vector is $M *$ `num_psd` by M. The frequency f_k in Hz associated with the PSD matrix k indexing is $f_k = (k - \text{num\_psd}/2)/(\text{num\_psd} * T_s)$ for $k = 1, \dots, \text{num\_psd}$.

```
function psd=minimum_variance_psd_MC(num_psd,p,T,x)

% This function computes the multichannel minimum variance spectral estimate
%
%           psd = minimun_variance_psd_MC(num_psd,p,T,x)
%
% num_psd -- number of frequency samples in PSD (must be power of two)
% p        -- order of minimum variance filter (number of taps = p + 1)
% T        -- sample interval in seconds
% x        -- sample data array:  x(sample #,channel #)
% psd      -- block vector of 'num_psd' multichannel PSD matrix elements

[num_pts,num_chs] = size(x);
n = 1:num_chs;
psd = zeros(num_chs*num_psd,num_chs);
I = eye(num_chs,num_chs);

% estimate multichannel AR parameter arrays using Nuttall-Strand multichannel AR
% algorithm; alternative choices are multichannel Yule-Walker or Vieira-Morf AR
[rho_a,a,rho_b,b] = lattice_MC('NS',p,x);
a = [I a];

% correlate the AR parameter arrays
psi = [ ];
for k=0:p
    pk = p + 1 - k;
    temp = pk*(rho_a\a(:,k*num_chs+n));
    for m=1:p-k
       temp = temp + (pk-m)*(a(:,m*num_chs+n)'/rho_a)*a(:,(k+m)*num_chs+n)...
                   - m*(b(:,(p-m-k)*num_chs+n)'/rho_b)*b(:,(p-m)*num_chs+n);
    end
    psi = [psi; temp];
end

% evaluate psi arrays with FFTs to yield min. var. PSD block vector
```

```
PSI = zeros(num_chs*num_psd,num_chs);
s = zeros(num_psd,1);
for k=n
    for m=k:num_chs
        s(1:p+1) = psi(m:num_chs:m+p*num_chs,k);
        s(num_psd:-1:num_psd-p+1) = conj(psi(k+num_chs:num_chs:k+p*num_chs,m));
        temp = fftshift(fft(s));
        PSI(m:num_chs:m+(num_psd-1)*num_chs,k) = temp;
        if m ~= k
            PSI(k:num_chs:k+(num_psd-1)*num_chs,m) = conj(temp);
        end
    end
end
for k=1:num_psd
    psd(n,:) = T*I/PSI(n,:);
    n = n + num_chs;
end
```

minimum_variance_2D

As described in *Digital Spectral Analysis* Section 16.9, the 2-D minimum variance spectral estimator takes the form

$$\mathbf{psd}_{\mathrm{MV}}(f1_k, f2_l) = \frac{T_1 T_2}{\underline{\mathbf{e}}^H(f1_k, f2_l)\mathbf{R}^{-1}\underline{\mathbf{e}}(f1_k, f2_l)}$$

in which $\underline{\mathbf{e}}(f1_k, f2_l)$ is the complex sinusoid block vector of Eq. (16.39) and $\mathbf{R}^{-1}$ is the inverse of the 2-D doubly-Toeplitz block autocorrelation matrix. The function output **psd** has dimension **num_psd1** by **num_psd2**. The row frequency indexing is $f_1(k) = (k - \mathbf{num\_psd1}/2)/(\mathbf{num\_psd1} * T_1)$ for $k = 1, \ldots, \mathbf{num\_psd1}$. The column frequency indexing $f_2(l) = (l - \mathbf{num\_psd2}/2)/(\mathbf{num\_psd2} * T_2)$ for $l = 1, \ldots, \mathbf{num\_psd2}$.

```
function psd=minimum_variance_psd_2D(num_psd1,num_psd2,p1,p2,T1,T2,x)
% Copyright (c) 2019 by S. Lawrence Marple Jr.

% This function computes the 2-D minimum variance spectral estimate
%
%     psd = minimum_variance_psd_2D(num_psd1,num_psd2,p1,p2,T1,T2,x)
%
% num_psd1 -- number of row frequency samples in  psd (must be power of 2)
% num_psd2 -- number of column frequency samples in psd (must be power of 2)
% p1   -- row order of minimum variance filter (number of row taps = p1+1)
% p2   -- column order of minimum variance filter (number of column taps = p2+1)
% T1   -- row sample interval (in seconds)
% T2   -- column sample interval (in seconds)
% x    -- sample data array of dimension N1 x N2
% psd -- first quadrant 2-D minimum variance spectrum (num_psd1 x num_psd2)

P1 = p1 + 1;
P2 = p2 + 1;
n = 1:P1;
I = eye(P1);

% Estimate 2-D quarter-plane reflection coefficient arrays using Marple lattice
% algorithm
[rho_f,a_f,rho_b,a_b,rho,A] = lattice_2D(p1,p2,x);
A = [I,A];

% Correlate the 2-D reflection parameter arrays
r = zeros(p1+1,2*p2+1)+1i*zeros(p1+1,2*p2+1);
for p = 0:p2
   pp = P2 - p;
   psi = pp*(rho\A(:,n+p*P1));
   for m = 1:p2-p
      psi = psi + (pp-m)*((A(:,n+m*P1)'/rho)*A(:,n+(m+p)*P1)) ...
        - m*(flipud(fliplr(conj( (A(:,n+(p2-m)*P1)'/rho)*A(:,n+(p2-m-p)*P1) ))));
   end
   % sum along diagonals of PSI to get correlation terms
   if p == 0
      psi = hermitian(psi);
```

```
      for m = 0:p1
         r(m+1,P2) = sum(diag(psi,m));
      end
   else
      for m = 0:p1
         r(m+1,P2+p) = sum(diag(psi,m));
         r(m+1,P2-p) = conj(sum(diag(psi,-m)));
      end
   end
end

% Use 2-D FFT on zero-padded correlation to evaluate the 2-D min.var. PSD
% p_inv = [1 zeros(1,p1)]/rho;
% rho_Q1 = 1/real(p_inv(1));
rr = zeros(num_psd1,num_psd2) + 1i*zeros(num_psd1,num_psd2);
rr(1:P1,1:P2) = r(:,P2:2*p2+1);
rr(1:P1,num_psd2-p2+1:num_psd2) = r(:,1:p2);
rr(num_psd1:-1:num_psd1-p1+1,num_psd2:-1:num_psd2-p2+1) = ...
    conj(r(2:P1,p2+2:2*p2+1));
rr(num_psd1:-1:num_psd1-p1+1,1:P2) = conj(r(2:P1,P2:-1:1));
psd = (T1*T2)*(ones(num_psd1,num_psd2)./fftshift(real(fft2(rr))));
```

modcovar_lp

This function solves, using a fast computational algorithm that exploits the Toeplitz infrastructure, the modified covariance method of linear prediction (LP) as described in *Digital Spectral Analysis* Section 8.5.2. The least squares minimization of the sum of both forward and backward LP results in the matrix solution

$$\mathbf{R}_p\mathbf{a}_p^{fb} = \begin{pmatrix} 2\rho_p \\ \mathbf{0}_p \end{pmatrix}$$

in which $\mathbf{0}_p$ is a p-element all-zeros vector,

$$\mathbf{R}_p = \begin{pmatrix} \mathbf{T}_p \\ \mathbf{T}_p^*\mathbf{J} \end{pmatrix}^H \begin{pmatrix} \mathbf{T}_p \\ \mathbf{T}_p^*\mathbf{J} \end{pmatrix} = \mathbf{T}_p^H\mathbf{T}_p + \mathbf{J}\mathbf{T}_p^T\mathbf{T}_p^*\mathbf{J}$$

and

$$\mathbf{T}_p = \begin{pmatrix} x[p+1] & \dots & x[1] \\ \vdots & \ddots & \vdots \\ x[N-p] & & x[p+1] \\ \vdots & \ddots & \vdots \\ x[N] & \dots & x[N-p] \end{pmatrix}.$$

The row vector a has dimension p.

```
function [rho,a]=modcovar_lp(p,x)
% Copyright (c) 2019 by S. Lawrence Marple Jr.

% Modified covariance least squares autoregressive parameter estimation
% algorithm using fast QR-decomposition linear prediction solution.
%
%     [rho,a] = modcovar_lp(p,x)
%
% p   -- order of linear prediction/autoregressive filter
% x   -- vector of data samples
% rho -- least squares estimate of linear prediction variance
% a   -- vector of linear prediction/autoregressive parameters

%*****************  Initialization  *****************

n = length(x);
if  3*p+1 > 2*n
    error(' Order too high; will make solution singular.')
end
a = [ ];
r1 = cabs2(x(2:n-1));
rho = r1 + cabs2(x(1)) + cabs2(x(n));
if p <= 0
    rho = rho/n;
    return
end
r2 = 1/(2*rho);
```

```
rho = rho + r1;
c = x(n)*r2;
d = conj(x(1))*r2;
ef = x;
eb = x;
ecf = c*x;
ecb = conj(c)*x;
edf = d*x;
edb = conj(d)*x;

%*************************  Main Recursion  *************************

for k=1:p

    if rho <= 0
        error(' Prediction squared error was less than or equal to zero.')
    end
    gam = 1 - real(ecb(n-k+1));
    del = 1 - real(edf(1));
    if (gam <=0 ) || (gam > 1) || (del <= 0) || (del > 1)
        error(' GAM or DEL gain factor not in expected range 0 to 1.')
    end

    % computation for k-th order reflection coefficient
    [eff,ef_k] = splitoff(ef,'top');
    [ebb,eb_n] = splitoff(eb,'bottom');
    k_p= -2*ebb'*eff/rho;

    % order update for squared prediction error  rho
    rho = rho*(1 - cabs2(k_p));

    % order update for linear prediction parameter array  a
    a = [a; 0] + k_p*[flipud(conj(a)); 1];

    % check if maximum order has been reached
    if k == p
        rho = .5*rho/(n-p);
        return
    end

    % order updates for prediction error arrays  ef and eb
    eb = ebb + conj(k_p)*eff;
    ef = eff + k_p*ebb;

    % coefficients for next set of updates
    c1 = ecf(1);                        % ecf(1) = edb(n-k+1) = conjg(lambda)
    c2 = edf(n-k+1);                    % edf(n-k+1) = conjg(ecb(1))
    c3 = ecf(n-k+1);
    c4 = conj(edb(1));
    r1 = 1/(gam*del - cabs2(c1));
    c5 = (c2*c1 + c3*del)*r1;
    c6 = (c3*conj(c1) + c2*gam)*r1;
    c7 = (c4*c1 + c2*del)*r1;
    c8 = (c2*conj(c1) + c4*gam)*r1;
```

```
    % time updates for gain arrays  c' and d'
    temp1 = c;
    c = c + c5*flipud(conj(c)) + c6*flipud(conj(d));
    d = d + c7*flipud(conj(temp1)) + c8*flipud(conj(d));
    % time updates for ecf', ecb', edf', and edb'
    temp1 = ecf;
    temp2 = ecb;
    temp3 = edf;
    ecf = ecf + c5*ecb + c6*edb;
    ecb = ecb + conj(c5)*temp1 + conj(c6)*edf;
    edf = edf + c7*temp2 + c8*edb;
    edb = edb + conj(c7)*temp1 +conj(c8)*temp3;
    [ecff,ecf_n] = splitoff(ecf,'bottom');
    [ecbb,ecb_k] = splitoff(ecb,'top');
    [edff,edf_n] = splitoff(edf,'bottom');
    [edbb,edb_k] = splitoff(edb,'top');

    if rho <= 0
        error(' Prediction squared error was less than or equal to zero.')
    end
    gam = 1 - real(ecbb(n-k));
    del = 1 - real(edff(1));
    if (gam <= 0) || (gam > 1) || (del <= 0) || (del > 1)
        error(' GAM or DEL gain factor not in expected range 0 to 1.')
    end

    % coefficients for next set of updates
    c1 = ecff(1);
    c2 = eb(n-k);
    c3 = ef(1);
    r1 = 1/(gam*del - cabs2(c1));
    c4 = (conj(c2)*del + c3*conj(c1))*r1;
    c5 = (c3*gam + conj(c2)*c1)*r1;
    c6 = c2/rho;
    c7 = conj(c3)/rho;

    % order updates for c and d; time update for a'
    temp1 = a;
    a = a + c4*c + c5*d;
    c = [0; c] + c6*[1; temp1];
    d = [0; d] + c7*[1; temp1];
    % time update for rho'
    rho = rho - real(c4*c2 + c5*conj(c3));

    % order updates for  ecf, ecb, edf, edb; time updates for  ef', eb'
    ecf = ecff + c6*ef;
    edf = edff +c7*ef;
    ef = ef +c4*ecff + c5*edff;
    ecb = ecbb + conj(c6)*eb;
    edb = edbb + conj(c7)*eb;
    eb = eb + conj(c4)*ecbb +conj(c5)*edbb;
end
```

noise_subspace

This function computes one of two noise subspace techniques. The MUSIC frequency estimator (not a true PSD)

$$\mathtt{psd}_{\mathrm{MUSIC}}(f_k) = \frac{1}{\mathbf{e}^H(f_k)\left(\sum_{k=M+1}^{p} \mathbf{v}_k \mathbf{v}_k^H\right)\mathbf{e}(f_k)},$$

based strictly on the noise subspace eigenvectors $\mathbf{v}$ with uniform weighting. Selecting an inverse eigenvalue weighting $1/\lambda_k$ yields the EV frequency estimator

$$\mathtt{psd}_{\mathrm{EV}}(f_k) = \frac{1}{\mathbf{e}^H(f)\left(\sum_{k=M+1}^{p} \frac{1}{\lambda_k}\mathbf{v}_k \mathbf{v}_k^H\right)\mathbf{e}(f)}.$$

The eigenvectors and eigenvalues are obtained from an SVD of the modified covariance data matrix [see *Digital Spectral Analysis* Section 13.6.2 for details]

$$\begin{pmatrix} x[p] & \dots & x[1] \\ \vdots & & \vdots \\ x[N-1] & \dots & x[N-p] \\ x^*[2] & \dots & x^*[p] \\ \vdots & & \vdots \\ x^*[N-p+1] & \dots & x^*[N] \end{pmatrix}.$$

The *psd* row vector has dimension $\mathtt{num\_psd}$ for which the frequency (Hz) is $f_k = (k - \mathtt{num\_psd}/2)/(\mathtt{num\_psd} * T)$ for $k = 1, \dots, \mathtt{num\_psd}$.

```
function psd=noise_subspace(num_psd,method,m,num_sig,x)
% Copyright (c) 2019 by S. Lawrence Marple Jr.

% This function computes spectral/frequency estimator by the MUSIC or EV method
% signal subspace analysis from the forward/backward data matrix.
%
%           psd = noise_subspace(num_psd,method,m,num_sig,x)
%
% num_psd -- number of spectral values to evaluate (power of two)
% method  -- select:  1--Min Norm method , 2--MUSIC algorithm , 3--EV algorithm
% m       -- dimension of data matrix ( prediction filter order )
% num_sig -- presumed number of complex sinusoidals in signal subspace
% x       -- vector of data samples
% psd     -- vector of  num_psd  frequency 'spectrum' values

if ( num_sig > m ) || ( num_sig < 0 )
    error([' Number of signals must be in range 0 <= num_sig <= ',int2str(m)])
end
```

```
n = length(x);
nm = num_sig+1:m+1;
X = toeplitz( x(m+1:n),x(m+1:-1:1) );
fb = [ X; fliplr(conj(X)) ];
[u,s,v] = svd(fb,0);

if  method == 1                            % Minimum Norm method
    g = v(1,nm);
    G = v(2:m+1,nm);
    temp = fft([1; (g*g')\G*g'],num_psd);
    psd = real(temp).^2 + imag(temp).^2;
else
    psd = zeros(num_psd,1);
    for k = nm
        temp = fft(v(:,k),num_psd);
        if  method == 2                    % MUSIC algorithm
            psd = psd + real(temp).^2 + imag(temp).^2;
        else                               % default: EV algorithm
            psd = psd + (1/s(k,k))*(real(temp).^2 + imag(temp).^2);
        end
    end
end
psd = fftshift(ones(num_psd,1)./psd);
```

periodogram_psd

This function computes the auto or cross periodogram using the Welch method (see *Digital Spectral Analysis* Section 5.7.3). The input signal column vector x (and vector y if a cross periodogram) is divided into segments of `seg_size` samples and overlap of `seg_overlap` samples between segments. The function will determine the number of segments J that can be created with the available signal record(s). For each segment j, an FFT is used to generate the Fourier transforms of the data samples $x^j[n]$ within that segment

$$X^j(f_k) = T \sum_{n=0}^{\texttt{seg\_size}-1} w[n]x^j[n]\exp(-j2\pi f_k nT)$$

$$Y^j(f_k) = T \sum_{n=0}^{\texttt{seg\_size}-1} w[n]y^j[n]\exp(-j2\pi f_k nT)$$

in which $w[n]$ is a sidelobe reducing window (three built in choices). The sample spectra of each segment j are then computed

$$P^j_{xx}(f_k) = \frac{1}{\text{seg\_size} * T} X^j(f_k)X^{j*}(f_k), \quad P^j_{xy}(f_k) = \frac{1}{\text{seg\_size} * T} X^j(f_k)Y^{j*}(f_k).$$

The segment sample spectra are then averaged to produce the final periodogram estimate:

$$\texttt{psd}_{xx}(f_k) = \frac{1}{J}\sum_{j=1}^{J} P^j_{xx}(f_k), \quad \texttt{psd}_{xy}(f_k) = \frac{1}{J}\sum_{j=1}^{J} P^j_{xy}(f_k).$$

Parameter T is the sampling interval (sec). The *psd* row vector has dimension `num_psd` for which the frequency (Hz) is $f_k = (k - \texttt{num\_psd}/2)/(\texttt{num\_psd} * T)$ for $k = 1, \ldots, \texttt{num\_psd}$.

```
function psd=periodogram_psd(num_psd,window,seg_overlap,seg_size,T,x,y)
% Copyright (c) 2019 by S. Lawrence Marple Jr.

% Classical periodogram auto/cross power spectral density estimate by Welch
% procedure.
%
%    Auto:   psd = periodogram_psd(num_psd,window,seg_overlap,seg_size,T,x)
%    Cross:  psd = periodogram_psd(num_psd,window,seg_overlap,seg_size,T,x,y)
%
% num_psd     -- number of psd vector elements (must be power of 2);
%                frequency spacing between elements is F = 1/(num_psd*T)
%                Hz, with psd(1) corresponding to frequency  -1/(2*T)
%                Hz, psd(num_psd/2 + 1) corresponding to  0  Hz, and
%                psd(num_psd) corresponding to 1/(2*T) - F  Hz.
% window      -- window selection: 0 -- none, 1 -- Hamming, 2 -- Nuttall
% seg_overlap -- number of overlap samples between segments
% seg_size    -- number of samples per segment (must be even)
```

```
% T            -- sample interval in seconds
% x            -- vector of data samples
% y            -- vector of data samples (cross PSD only);
%                  >>note that length(y) must equal length(x)<<
% psd          -- vector of num_psd auto/cross PSD values

if nargin == 7
    if length(x) ~= length(y)
        error(' The input data vectors are not of equal lengths.')
    end
end
if seg_size > length(x)
    error([' Seg_size cannot exceed data record = ',int2str(length(x))])
end
if seg_size > num_psd
    error([' Seg_size cannot exceed number of PSD freqs = ',int2str(num_psd)])
end
if (seg_overlap < 0) || (seg_overlap >= seg_size)
    error([' Out of range: 0 <= overlap <',int2str(seg_size)])
end
shift = seg_size - seg_overlap;
num_segs = fix(( length(x) - seg_size )/shift) + 1;
psd = zeros(num_psd,1);
range = 1:seg_size;
s = seg_size - 1;
ph = 2*pi*(0:s)/s;
if     window == 0  wind = ones(seg_size,1);           % rectangular (aka boxcar)
elseif window == 1  wind = (.53836 - .46164*cos(ph))';                  % Hamming
elseif window == 2  wind=(.42323-.49755*cos(ph)+.07922*cos(2*ph))';     % Nuttall
else
    error(' Window selection number is invalid.')
end
% Window could also be specified as one of the following from the MATLAB Signal
% Processing Toolbox:
%   wind = barlett(seg_size);
%   wind = blackman(seg_size);
%   wind = chebwin(seg_size,sidelobe level in db);
%   wind = hanning(seg_size);
%   wind = kaiser(seg_size,0.1102*(sidelobe level in db - 8.7));
%   wind = triang(seg_size);

for k=1:num_segs
    z = T*fft(wind.*x(range),num_psd);
    if nargin < 7
        psd = psd + real(z).^2 + imag(z).^2;
    else
        psd = psd + z.*(T*conj(fft(wind.*y(range),num_psd)));
    end
    range = range + shift;
end
pow_wind = sum(wind.^2)/seg_size;          % adjustment in gain for window power
psd = (1/(num_segs*pow_wind*seg_size*T))*fftshift(psd);
```

periodogram_psd_MC

This function computes the multichannel version of the periodogam power spectral density estimator (see *Digital Spectral Analysis* Section 15.5). A multichannel signal matrix $\mathbf{x}(n, m)$ of dimension N row samples by M column signal channels is provided as an input from which the dimension $M \times M$ matrix

$$\mathbf{psd}(f_k) = \begin{pmatrix} \text{psd}_{11}(f_k) & \text{psd}_{12}(f_k) & \dots & \text{psd}_{1M}(f_k) \\ \text{psd}_{21}(f_k) & \text{psd}_{22}(f_k) & \dots & \text{psd}_{2M}(f_k) \\ \vdots & \vdots & \ddots & \vdots \\ \text{psd}_{M1}(f_k) & \text{psd}_{M2}(f_k) & \dots & \text{psd}_{MM}(f_k) \end{pmatrix}$$

is formed at each frequency f_k by populating the matrix with auto and cross psd values generated by auto and cross psd estimates provided by `periodogram_psd`. Each individual frequency **psd** matrix is stacked to form a block column vector of `num_psd` elements. The corresponding scalar dimension of this block column vector is $M *$ `num_psd` by M. The frequency f_k in Hz associated with the **psd** matrix k indexing is $f_k = (k - \texttt{num\_psd}/2)/(\texttt{num\_psd} * T)$ for $k = 1, \dots, \texttt{num\_psd}$. See function `periodogram_psd` for a description of the function input parameters `window`, `overlap`, `seg_size` and `T`.

```
function psd=periodogram_psd_MC(num_psd,window,overlap,seg_size,T,x)
% Copyright (c) 2019 by S. Lawrence Marple Jr.

% This function computes the classical multichannel periodogram power spectral
% density matrix over 'num_psd' frequencies.
%
%          psd = periodogram_psd_MC(num_psd,window,overlap,seg_size,T,x)
%
% num_psd  -- number of frequency points in psd (must be a power of two)
% window   -- window selection:  1 -- none , 1 -- Hamming , 2 -- Nuttall
% overlap  -- number of overlap samples between segments
% seg_size -- number of samples per segment (must be even)
% T        -- sample interval in seconds
% x        -- sample data array:  x(sample #,channel #)
% psd      -- block vector of 'num_psd' psd matrices

[num_pts,num_chs] = size(x);
k = 0:num_chs:(num_psd - 1)*num_chs;

for m=1:num_chs
    psd(k+m,m) = periodogram_psd(num_psd,window,overlap,seg_size,T,x(:,m));
    for n=m+1:num_chs
        power = periodogram_psd(num_psd,window,overlap,seg_size,T,x(:,m),x(:,n));
        psd(k+m,n) = power;
        psd(k+n,m) = conj(power);
    end
end
```

periodogram_psd_2D

This function computes a 2-D periodogram spectral estimate. Assuming a 2-D signal array of M by N samples, a nonaveraged periodogram of the entire array is calculated as

$$P_{\text{PER}}(f1_k, f2_l) = \frac{T_1 T_2}{MN} \left| \sum_{m=0}^{M-1} \sum_{n=0}^{N-1} w[m,n] x[m,n] \exp(-j2\pi[f1_k m T_1 + f2_l n T_2]) \right|^2 ,$$

in which $w[m,n]$ is an appropriate data window. The 2-D signal array can be broken into smaller subarrays and the 2-D sample periodogram computed of each subarray. These subarray periodograms can be averaged to produce the final 2-D **psd**. See Chapter 5 of this user guide for how the subarrays are specified. The function output **psd** has dimension num_psd1 by num_psd2. The row frequency indexing is $f_1(k) = (k - \text{num\_psd1}/2)/(\text{num\_psd1} * T_1)$ for $k = 1, \ldots, \text{num\_psd1}$. The column frequency indexing $f_2(l) = (l - \text{num\_psd2}/2)/(\text{num\_psd2} * T_2)$ for $l = 1, \ldots, \text{num\_psd2}$.

```
function psd=periodogram_psd_2D(num_psd1,num_psd2,window,overlap1,overlap2,...
                                             seg_size1,seg_size2,T1,T2,x)
% Copyright (c) 2019 by S. Lawrence Marple Jr.

% Classical two-dimensional periodogram power spectral density estimate by
% subarray averaging procedure of equation (16.40).
%
%    psd = periodogram_psd_2D(num_psd1,num_psd2,window,overlap1,overlap2,...
%                                            seg_size1,seg_size2,T1,T2,x)
%
% num_psd1   -- number of row PSD values to evaluate (must be power of two)
% num_psd2   -- number of column PSD values to evaluate (must be power of two)
% window     -- window selection:  0 [none], 1 [Hamming], 3 [Nutalll]
% overlap1   -- subarray overlap in row dimension (# samples)
% overlap2   -- subarray overlap in column dimension (# samples)
% seg_size1 -- row dimension of data subarray segment (# samples)
% seg_size2 -- column dimension of data subarray segment (# samples)
% T1         -- sample interval along rows in seconds
% T2         -- sample interval along columns in seconds
% x          -- two-dimensional sample data array
% psd        -- two dimensional power spectral density array

[n1,n2] = size(x);
if seg_size1 > n1
    error([' Seg_size1 cannot exceed data row dimension = ',int2str(n1)])
end
if seg_size2 > n2
    error([' Seg_size2 cannot exceed data column dimension = ',int2str(n2)])
end
if (overlap1 < 0) || (overlap1 >= seg_size1)
    error([' Out of range; must select 0 < overlap1 < ',int2str(seg_size1)])
end
if (overlap2 < 0) || (overlap2 >= seg_size2)
```

```
    error([' Out of range; must select 0 < overlap2 < ',int2str(seg_size2)])
end

s1 = seg_size1 - 1;
s2 = seg_size2 - 1;
ph1 = 2*pi*(0:s1)/s1;
ph2 = 2*pi*(0:s2)/s2;
if window == 0
    wind = ones(seg_size1,seg_size2); % rectangular (aka boxcar)
elseif window == 1
    wind = (.53836-.46164*cos(ph2))'*(.53836-.46164*cos(ph1)); % Hamming
elseif window == 2
    wind = (.42323-.49755*cos(ph2)+.07922*cos(2*ph2))'*...
           (.42323-.49755*cos(ph1)+.07922*cos(2*ph1)); % Nuttall
else
    error(' Window selection number is invalid.')
end

shift1 = seg_size1 - overlap1;
shift2 = seg_size2 - overlap2;
num_ints1 = fix((n1 - seg_size1)/shift1) +1;
num_ints2 = fix((n2 - seg_size2)/shift2) +1;
psd = zeros(num_psd1,num_psd2);

range1 = 1:seg_size1;
for k1 = 1:num_ints1
    range2 = 1:seg_size2;
    for k2 = 1:num_ints2
        xfm = fft2(wind.*x(range1,range2),num_psd1,num_psd2);
        psd = psd + real(xfm).^2 + imag(xfm).^2;
        range2 = range2 + shift2;
    end
    range1 = range1 + shift1;
end

if (T1 > 0) && (T2 > 0)
    psd = fftshift(psd)*(T1*T2/(num_ints1*num_ints2*seg_size1*seg_size2));
else
    psd = fftshift(psd);
end
```

pisarenko

This function implements the five steps of the Pisarenko technique to fit a sum of sinusoidals plus white noise model to a signal data record, described in *Digital Spectral Analysis* Section 13.6.1. These are:

- Estimate the autocorrelation sequence $r[m]$ from the input data and form the Toeplitz autocorrelation matrix $\mathbf{R}$ from the $r[0]$ to $r[\texttt{num\_sig}]$ estimates.
- Solve the eigenequation $\mathbf{R}\mathbf{v} = \lambda\mathbf{v}$ for the minimum eigenvalue λ and its associated eigenvector $\mathbf{v}$. If the model is appropriate to the signal, the minimum eigenvalue will be the white noise variance.
- Use the eigenvector parameters to form a polynomial, which is factored to obtain the sinusoidal frequencies f_k for $k = 1, \ldots, \texttt{num\_sigs}$.
- Solve the following linear equation model relationship between the autocorrelation values and the sinusoidal frequencies/powers to estimate the sinusoidal powers

$$r_{xx}[k] = \sum_{i=1}^{M} P_i \cos(2\pi f_i kT) + \rho_w \delta[k].$$

A special **psd** column vector of dimension `num_psd` places a flat line across all frequencies to represent white noise level, and single spectral lines at the estimated sinusoidal frequencies at relative levels that represent their powers.

```
function [rho,freqs,powers,psd]=pisarenko(num_psd,num_sig,T,x)
% Copyright (c) 2019 by S. Lawrence Marple Jr.

% Pisarenko harmonic decomposition frequency/spectral estimator.
%
%     [rho,freqs,powers,psd] = pisarenko(num_psd,num_sig,T,x)
%
% num_psd -- number of spectral points for plot
% num_sig -- number of presumed complex sinusoidals (need 2 for each real sine)
% T       -- sample interval in seconds
% x       -- vector of data samples
% rho     -- white noise power spectral density
% freqs   -- vector of estimated sinusoid frequencies
% powers  -- vector of estimated sinusoid powers
% psd     -- vector of num_psd spectral points

% estimate biased autocorrelation sequence (ACS)
r = correlation_sequence(num_sig,'biased',x);

% solve for minimum eigenvalue ( rho is white noise variance estimate )
[rho,eigvec] = minimum_eigenvalue( r(num_sig+1:2*num_sig+1) );
```

```
% estimate complex sinusoid frequencies ( in Hertz )
z = roots(eigvec);
freqs = angle(z)/(2*pi*T);

% estimate complex sinusoid powers
powers = abs(vandermonde_lineqs(z,r(num_sig+2:2*num_sig+1)));

% create novel Pisarenko line spectrum embedded in constant white noise PSD
rho = rho*T;
psd = rho*ones(num_psd,1);      % white noise power spectral density level
for k=1:num_sig
    psd(fix( num_psd*(freqs(k)*T + .5) )) = powers(k);
end
```

symcovar

This function implements the symmetric covariance case of linear prediction (LP) used in the Prony method, as described in *Digital Spectral Analysis* Section 11.6. This technique forces a symmetric LP parameter fit to the signal data, which results in undamped sinusoidal frequencies when used in the polynomial factoring step of the least squares Prony method. The structure of the matrix expression to be solved is

$$\mathbf{R}_{2p}\mathbf{g}_{2p} = \begin{pmatrix} \mathbf{0}_p \\ 2\rho_p^s \\ \mathbf{0}_p \end{pmatrix},$$

in which the centrosymmetric matrix $\mathbf{R}_{2p}$ and conjugate symmetric vector $\mathbf{g}_{2p}$ are

$$\mathbf{R}_{2p} = \begin{pmatrix} r_{2p}[0,0] & \dots & r_{2p}[0,2p] \\ \vdots & & \vdots \\ r_{2p}[2p,0] & \dots & r_{2p}[2p,2p] \end{pmatrix}, \quad \mathbf{g}_{2p} = \begin{pmatrix} g_p[p] \\ \vdots \\ g_p[1] \\ 1 \\ g_p^*[1] \\ \vdots \\ g_p^*[p] \end{pmatrix}.$$

The matrix $\mathbf{R}$ has a Toeplitz structure that may be exploited to develop a fast computational algorithm that is listed below. All input/output parameters are column vectors.

```
function [rho_s,g]=symcovar(p,x)
% Copyright (c) 2019 by S. Lawrence Marple Jr.

% Fast algorithm for the solution of the hermitian symmetric covariance least
% squares normal equations.
%
%          [rho_s,g] = symcovar(p,x)
%
% p     -- order of linear smoothing filter (must be even)
% x     -- vector of data samples
% rho_s -- least squares estimate of linear smoothing variance
% g     -- vector of hermitian symmetric linear smoothing parameters

%************************  Initialization  ************************

if rem(p,2) ~= 0
    error(' Order for SYMCOVAR must be even.')
end
n = length(x);
a = [ ];
r1 = cabs2(x(2:n-1));
rho = r1 + cabs2(x(1)) + cabs2(x(n));
if  p <= 0
    rho_s = rho/n;
```

```
    return
end
g = 1/(2*rho);
r2 = g;
rho = rho + r1;
c = x(n)*r2;
d = conj(x(1))*r2;
ef = x;
eb = x;
ecf = c*x;
ecb = conj(c)*x;
edf = d*x;
edb = conj(d)*x;

%************************  Main Recursion  ************************

for k=1:p

    if  rho <= 0
        error(' Prediction squared error ''rho'' <= 0 .')
    end
    gam = 1 - real(ecb(n-k+1));
    del = 1 - real(edf(1));
    if (gam <=0 ) || (gam > 1) || (del <= 0) || (del > 1)
        error(' Gain factor ''gam'' or ''del'' not in range 0 to 1 .')
    end

    % update k-th order reflection coefficient  k_p
    [eff,ef_k] = splitoff(ef,'top');
    [ebb,eb_n] = splitoff(eb,'bottom');
    k_p= -2*ebb'*eff/rho;

    % order update for squared prediction error  rho
    rho = rho*(1 - cabs2(k_p));
    if rem(k,2) == 0
        c1 = a(k/2);
    end

    % order update for autoregressive/linear prediction parameter array  a
    a = [a; 0] + k_p*[flipud(conj(a)); 1];

    % order update of linear smoothing parameter array  g
    if rem(k,2) == 0
        c1 = c1/rho;
        g = [0;g;0] + conj(c1)*[1;a] + c1*[flipud(conj(a));1];
    end

    % determine if final order has been reached
    if k == p
        c1 = 1/real(g(p/2+1));
        rho_s = c1/(2*(n-p));
        g = c1*g;
        return
    end
```

```
% order updates for prediction error arrays  ef and eb
ef = eff + k_p*ebb;
eb = ebb + conj(k_p)*eff;

% precomputation of constants for next set of updates
c1 = ecf(1);                % note:  ecf(1) = edb(n-k+1) = conjg(lambda)
c2 = edf(n-k+1);            % note:  edf(n-k+1) = conjg(ecb(1))
c3 = ecf(n-k+1);
c4 = conj(edb(1));
r1 = 1/(gam*del - cabs2(c1));
c5 = (c2*c1 + c3*del)*r1;
c6 = (c3*conj(c1) + c2*gam)*r1;
c7 = (c4*c1 + c2*del)*r1;
c8 = (c2*conj(c1) + c4*gam)*r1;

% time update of linear smoothing parameter array  g"
if rem(k,2) ~= 0
    kh = (k-1)/2 + 1;
    bb = [conj(d(kh)); d(kh); conj(c(kh)); c(kh)];
    aa = [ del -conj(c4) -c1 -conj(c2);
            0     del    -c2 -conj(c1);
            0      0     gam -conj(c3);
            0      0      0        gam; ];
    rr = chol(aa);
    y = rr'\bb;
    bb = rr\y;
    g = g + bb(3)*c + bb(4)*flipud(conj(c)) + bb(1)*d+bb(2)*flipud(conj(d));
end

% time updates for gain arrays  c' and d'
temp1 = c;
c = c + c5*flipud(conj(c)) + c6*flipud(conj(d));
d = d + c7*flipud(conj(temp1)) + c8*flipud(conj(d));

% time updates for  ecf', ecb', edf', and edb'
temp1 = ecf;
temp2 = ecb;
temp3 = edf;
ecf = ecf + c5*ecb + c6*edb;
ecb = ecb + conj(c5)*temp1 + conj(c6)*edf;
edf = edf + c7*temp2 + c8*edb;
edb = edb + conj(c7)*temp1 +conj(c8)*temp3;
[ecff,ecf_n] = splitoff(ecf,'bottom');
[ecbb,ecb_k] = splitoff(ecb,'top');
[edff,edf_n] = splitoff(edf,'bottom');
[edbb,edb_k] = splitoff(edb,'top');

if rho <= 0
    error(' Prediction squared error ''rho'' <= 0 .')
end
gam = 1 - real(ecbb(n-k));
del = 1 - real(edff(1));
if (gam <= 0) || (gam > 1) || (del <= 0) || (del > 1)
    error(' Gain factor ''gam'' or ''not in range 0 to 1 .')
end
```

```
    % precomputation of constants for next set of updates
    c1 = ecff(1);
    c2 = eb(n-k);
    c3 = ef(1);
    r1 = 1/(gam*del - cabs2(c1));
    c4 = (conj(c2)*del + c3*conj(c1))*r1;
    c5 = (c3*gam + conj(c2)*c1)*r1;
    c6 = c2/rho;
    c7 = conj(c3)/rho;

    % order updates for  c and d; time update for  a'
    temp1 = a;
    a = a + c4*c + c5*d;
    c = [0; c] + c6*[1; temp1];
    d = [0; d] + c7*[1; temp1];

    % time update for prediction squared error  rho'
    rho = rho - real(c4*c2+c5*conj(c3));

    % order updates for  ecf, ecb, edf, edb; time updates for  ef', eb'
    ecf = ecff + c6*ef;
    edf = edff +c7*ef;
    ef = ef +c4*ecff + c5*edff;
    ecb = ecbb + conj(c6)*eb;
    edb = edbb + conj(c7)*eb;
    eb = eb + conj(c4)*ecbb +conj(c5)*edbb;
end
```

toeplitz_lineqs

This function solves a set of $p+1$ linear equations (complex-valued or real-valued) involving a Toeplitz matrix (see *Digital Spectral Analysis* Section 3.8.3)

$$\mathbf{T}_p \mathbf{x}_p = \mathbf{z}_p,$$

in which $\mathbf{z}_p$ is a known right-hand side $(p+1)$-dimensional column vector, and

$$\mathbf{T}_M = \begin{pmatrix} t[0] & t[-1] & \cdots & t[-p] \\ t[1] & t[0] & \cdots & t[p-1] \\ \vdots & \vdots & \ddots & \vdots \\ t[p] & t[p-1] & \cdots & t[0] \end{pmatrix}$$

is the Toeplitz matrix. The matrix is not explicitly needed as the left-hand side column vector and top row vector contain all the elements needed to form $\mathbf{T}$. Exploitation of the structure of the matrix results in a fast algorithm requiring computations proportional to p^2 rather than the normal p^3 computations for linear equations solution.

```
function x=toeplitz_lineqs(T_left,T_top,y)
% Copyright (c) 2019 by S. Lawrence Marple Jr.

% Finds linear equation solution for nonsymmetric Toeplitz matrix: Tx = y
%
%            x = toeplitz_lineqs(T_left,T_top,y)
%
% T_left -- left column of Toeplitz matrix (all elements generally complex)
% T_top  -- top row of Toeplitz matrix (all elements generally complex-valued)
% y      -- given y column vector (generally complex-valued)
% x      -- solution column vector (can be complex-valued)

%*******************  Initialization and Memory Allocation  *****************

if T_top(1) ~= T_left(1)
    error(' T_left(1) does not equal T_top(1).')
end
N = length(T_top);
if N ~= length(T_left)
    error(' Dimension of top row and left column are not equal.')
end
x(1) = y(1)/T_top(1);

%*********************************  Main Recursion  ************************

for k=1:N-1

        % estimate partial correlation matrix delta
        if k > 1
            delta_a = [a 1]*T_left(2:k+1);
            delta_b = [1 b]*flipud(T_top(2:k+1).');
```

```
        delta_c = x*flipud(T_top(2:k+1).');
    else    % for order 1
        delta_a = T_left(2);
        delta_b = T_top(2);
        delta_c = x(1)*T_top(2);
        p = T_left(1);
    end

    % update forward and backward reflection coefficient matrices
    ka = -delta_a/p;
    kb = -delta_b/p;

    % update forward and backward error covariance matrices
    p = (1 - ka*kb)*p;

    % update forward and backward prediction parameter block vectors
    if k > 1
        a_save = a;
        a = [0 a_save] + ka*[1 b];
        b = [b 0] + kb*[a_save 1];
    else    % for order 1
        a = ka;
        b = kb;
    end
    x = [x 0] + ((y(k+1)-delta_c)/p)*[a 1];

end
```

vandermonde_lineqs

This function solves a set of p linear equations (complex-valued or real-valued) involving a Vandermonde matrix (see *Digital Spectral Analysis* Section 3.9)

$$\mathbf{V}_p \mathbf{x}_p = \mathbf{y}_p,$$

in which $\mathbf{y}_p$ is a known right-hand side $(p+1)$-dimensional column vector, and

$$\mathbf{V} = \begin{pmatrix} v_1^0 & v_2^0 & \cdots & v_p^0 \\ \vdots & \vdots & & \vdots \\ v_1^{p-1} & v_2^{p-1} & \cdots & v_p^{p-1} \end{pmatrix}$$

is the Vandermonde matrix. The matrix is not explicitly needed as only the elements v_k for $k = 1, \ldots, p$ in vector $\mathbf{v}$ are needed to form $\mathbf{V}$. Exploitation of the structure of the matrix results in a fast algorithm requiring computations proportional to p^2 rather than the normal p^3 computations for linear equations solutions.

```
function x=vandermonde_lineqs(v,y)
% Copyright (c) 2019 by S. Lawrence Marple Jr.

% Solves the complex linear simultaneous equations  Vx = y  in which  V is a
% Vandermonde matrix using the fast computational algorithm of Bjorck and
% Pereyra (Section 3.9).
%
%     x = vandermonde_lineqs(v,y)
%
% v -- vector or complex Vandermonde basis elements
% y -- right-hand-side vector with complex elements
% x -- solution vector

x = y;
m = length(v);
for k = 1:m-1
    z = x(k:m-1);
    x(k+1:m) = x(k+1:m) - v(k)*z;
end
for k = m-1:-1:1
    for j = k+1:m
        temp = v(j) - v(j-k);
        if temp == 0
            error(' Two Vandermonde elements cannot be equal.')
        end
        x(j) = x(j)/temp;
    end
    for j = k:m-1
        x(j) = x(j) - x(j+1);
    end
end
```

yule_walker

This function implements the Yule-Walker approach to estimating autoregressive (AR) parameters and the associated white noise variance from a single signal data vector. It pairs functions `correlation_sequence` and `levinson_recursion` to accomplish this estimate (see *Digital Spectral Analysis* Section 8.3).

```
function [rho,a]=yule_walker(p,x)

% Yule-Walker autoregressive parameter estimation algorithm.
%
%     [rho,a] = yule_walker(p,x)
%
% p   -- order of autoregressive (AR) process
% x   -- vector of data samples
% rho -- noise variance estimate
% a   -- vector of autoregressive parameters

r = correlation_sequence(p,'biased',x);
[rho,a] = levinson_recursion(r(p+1:2*p+1));
```

yule_walker_MC

This function implements an extension of the Yule-Walker approach to estimating multi-channel autoregressive (AR) parameter matrices and the associated white noise variance matrix from a multichannel signal data array (samples along rows and channels along columns). It pairs functions `correlation_sequence_MC` and `levinson_recursion_MC` to accomplish this estimate (see *Digital Spectral Analysis* Section 15.10.1).

```
function [rho_a,a,rho_b,b]=yule_walker_MC(p,x)
% Copyright (c) 2019 by S. Lawrence Marple Jr.

% Multichannel Yule-Walker autoregressive parameter estimation algorithm.
%
%           [rho_a,a,rho_b,b] = yule_walker_MC(p,x)
%
% p      -- order of multichannel autoregressive (AR) process
% x      -- sample data matrix:  x(sample #,channel #)
% rho_a -- forward linear prediction error/white noise covariance matrix
% a      -- forward linear prediction/autoregressive (AR) parameter block vector
% rho_b -- backward linear prediction error/white noise covariance matrix
% b      -- backward linear prediction/autoregressive (AR) parameter block vector

r = correlation_sequence_MC(p,'biased',x);
[rho_a,a,rho_b,b] = levinson_recursion_MC(r);
```

yule_walker_2D

This function implements an extension of the Yule-Walker approach to estimating 2-D Q1 and Q4 quadrants autoregressive (AR) parameter matrices and the associated quadrant white noise variances from a 2-D signal data array. Pairing functions `correlation_sequence_2D` and `levinson_recursion_2D` will accomplish this estimate (see *Digital Spectral Analysis* Sections 16.5, 16.7.3, and 16.7.4).

```
function [rho_Q1,a_Q1,rho_Q4,a_Q4]=yule_walker_2D(p1,p2,x)
% Copyright (c) 2019 by S. Lawrence Marple Jr.

% Two-dimensional Yule-Walker autoregressive parameter estimation algorithm.
%
%     [rho_Q1,a_Q1,rho_Q4,a_Q4] = yule_walker_2D(p1,p2,x)
%
% p1     -- row order of 2-D autoregressive (AR) process
% p2     -- column order of 2-D autoregressive (AR) process
% x      -- sample data array:  x(row sample #,column sample #)
% rho_Q1 -- first quadrant (Q1) white noise variance
% a_Q1   -- first quadrant (Q1) 2-D autoregressive parameter array
% rho_Q4 -- fourth quadrant (Q4) white noise variance
% a_Q4   -- fourth quadrant (Q4) 2-D autoregressive parameter array

r = correlation_sequence_2D(p1,p2,x);
[rho_Q1,a_Q1,rho_Q4,a_Q4] = levinson_recursion_2D(r);
```